Digital Applications for CPLDs:
A Lab Manual

Robert K. Dueck
Red River College

Delmar Publishers

an International Thomson Publishing company I(T)P®

Albany · Bonn · Boston · Cincinnati · Detroit · London · Madrid · Melbourne
Mexico City · New York · Pacific Grove · Paris · San Francisco · Singapore
Tokyo · Toronto · Washington

NOTICE TO THE READER

Delmar Staff:
Publisher: Michael McDermott
Acquisitions Editor: Gregory Clayton
Developmental Editor: Michelle Ruelos Cannistraci
Production Manager: Larry Main
Art and Design Coordinator: Nicole Reamer
Marketing Coordinator: Paula Collins
Editorial Assistant: Amy Tucker

COPYRIGHT © 2000
By Delmar Publishers
a division of International Thomson Publishing

The ITP logo is a trademark under license.

Printed in the United States of America

For more information, contact:
Delmar Publishers
3 Columbia Circle, Box 15015
Albany, New York 12212-5015

International Thomson Publishing Europe
Berkshire House
168-173 High Holborn
London WC1V 7AA
United Kingdom

Nelson ITP, Australia
102 Dodds Street
South Melbourne
Victoria, 3205 Australia

Nelson Canada
1120 Birchmont Road
Scarborough, Ontario
Canada, M1K 5G4

International Thomson Publishing France
Tour Maine-Montparnasse
33 Avenue du Maine
75755 Paris Cedex 15, France

International Thomson Editores
Seneca 53
Colonia Polanco
11560 Mexico D. F. Mexico

International Thomson Publishing GmbH
Königswinterer Strasse 418
53227 Bonn
Germany

International Thomson Publishing Asia
60 Albert Street
#15-01 Albert Complex
Singapore 189969

International Thomson Publishing- -Japan
Hirakawacho Kyowa Building, 3F
2-2-1 Hirakawacho
Chiyoda-ku, Tokyo 102
Japan

ITE Spain/Paraninfo
Calle Magallanes, 25
28015-Madrid, Espana

Online Services

Delmar Online
To access a wide variety of Delmar products and services on the World Wide Web, point your browser to:
http://www.delmar.com/delmar.html
or email: info@delmar.com

thomson.com
To access International Thomson Publishing's home site for information on more than 34 publishers and 20,000 products, point your browser to:
http://www.thomson.com
or email: findit@kiosk.thomson.com

A service of I(T)P®

2 3 4 5 6 7 8 9 10   XXX   03 02 01 00

ISBN 0-7668-1696-6

Contents

Intended Audience

This series of labs is intended for use in a digital design sequence at the Computer/Electronics Engineering Technology (CET/EET) level.

Digital Design: A New Paradigm

Until now, digital logic or digital design courses at the EET level have primarily focussed on using fixed-function TTL and CMOS integrated circuits as the vehicle for teaching principles of logic design. However, the digital design field has turned a corner; more and more, digital designs are being implemented in Programmable Logic Devices (PLDs), rendering many of the popular fixed-function devices obsolete.

This lab series endeavors to address this new trend by focussing solely on PLDs as a vehicle for teaching the new digital paradigm. The labs use Altera's MAX+PLUS II PLD design and programming software in its Student Edition and the Altera UP-1 University Program Laboratory Design Package as the hardware platform.

The Altera UP-1 board has two Complex PLDs (CPLDs): one from the MAX 7000S family (EPM7128SLC84-7) and one from the FLEX10K family (EPF10K20RC240-4). The EPM7128S is a Sum-of-Products (SOP) device with 128 logic cells, based on EEPROM technology (and therefore nonvolatile). The EPF10K20 is a Look-Up Table (LUT) device, based on SRAM cells (and therefore volatile). Either or both chips can be programmed or configured without removing them from the circuit. This set of labs uses the EPM7128S chip exclusively. Although these labs are geared to the Altera UP-1 board, any board with an EPM7128S CPLD and suitable input/output devices can be used.

The labs make use of the various design entry and simulation features of MAX+PLUS II, including schematic capture, text entry using VHDL (VHSIC Hardware Description Language; VHSIC = Very High Speed Integrated Circuit), and use of components from the Library of Parameterized Modules (LPM). The labs are stepped in difficulty so that a student initially is asked only to follow procedures without making too many decisions and in later labs is given broad directions, requiring some design decision-making. The principles taught in Labs 1 through 5 are brought together in a design project in Lab 6. Lab 7 is a more advanced unit (interfacing to a digital-to-analog converter) that stands on its own.

In addition there is a tutorial on MAX+PLUS II design entry and programming, a tutorial on VHDL basics, and a design case study that shows VHDL implementation of a switch debouncer, using LPM components.

Outline of Units and Objectives

• Tutorial 1   Programming CPLDs using MAX+PLUS II

Introduction to design entry (schematic capture and VHDL and programming of a CPLD, using MAX+PLUS II. *It is strongly recommended that you work along with the tutorial examples in MAX+PLUS II before proceeding with the remainder of the lab material.*

- ## Tutorial 2   Introduction to VHDL

An outline of basic VHDL structures, such as entity declaration, architecture body, data types, libraries, concurrent and selected signal assignments, PROCESS and CASE statements. The tutorial uses combinational circuits (decoder and multiplexer) and sequential circuits (counters) as examples.

- ## Lab 1   Introduction to the Altera UP-1 Board

Download a Programmer Object File (pof) describing a 2-bit magnitude comparator to a CPLD on the Altera UP-1 board. Wire the chip to input switches and output LEDs on the UP-1 board and demonstrate circuit operation. No design entry is required in this unit.

- ## Lab 2   Seven-Segment Decoder

Type in a VHDL file and complete a truth table represented in a selected signal assignment statement. (Most of the VHDL file is given.) Create a decoder symbol from the VHDL file for use in a Graphic Design File. Connect two instances (copies) of the decoder to an 8-bit counter (predesigned) to demonstrate a two-digit hexadecimal count on the seven-segment displays on the Altera UP-1 board.

- ## Lab 3   Multiplexer Applications

Create a variety of multiplexer circuits using schematic capture and VHDL. Create simulations of each design to verify their operation. Program a CPLD with each multiplexer circuit (digital signal selector, pattern generator, hexadecimal digit selector.)

- ## Lab 4   Counters and Decoders

Enter the design for a 3-bit synchronous counter (schematic capture). Create a simulation for the counter and incorporate it into a design with a seven-segment decoder and a binary decoder. Create a simulation for the counter/binary decoder combination. Download and test these designs.

- ## Lab 5   Parameterized Counters and Shift Registers

Enter, simulate, and test designs for counters and shift registers using components from the Library of Parameterized Modules (LPM).

- ## Lab 6   Time-Division Multiplexing (Design Project)

Use knowledge from Labs 1 through 5 to create a system that synchronously transfers four 4-bit words between two Altera UP-1 boards across an interface consisting of two transmit/receive data lines (full duplex), a clock, a reset, and a ground line.

- ## Lab 7   DAC Function Generator

Create an interface to a digital-to-analog converter that allows manual input to the DAC as well as the ability to generate sawtooth, square, and triangle waves.

- ## Design Case Study   Switch Debouncer for the Altera UP-1 Board

Shows the design of a switch debouncer, based on a 16-bit clock divider and a 4-bit shift register, using VHDL and LPM components.

Class Time Requirements

The labs in this package are generally longer than digital labs in most of the current EET lab manuals available on the market. With the exception of Labs 1 and 2, each of which could probably be completed in a single lab session, these labs are designed for about 4–6 lab hours each. The design project in Lab 6 requires about 10–12 hours.

In traditional TTL/breadboard labs, this would be problematic due to the time required for circuit connection and the inevitable task of troubleshooting that accompanies it. However, when programming PLDs with MAX+PLUS II, the majority of design setup is done in software. This lends a flexibility and portability that has not been available until now. Students can work on their labs on their own time, at home or in a computer lab, requiring the hardware only for final design testing and demonstration.

Also, the CPLD pin assignments for Labs 1 through 6 are standard; it is possible to wire the switch/LED connections on the Altera UP-1 board once and then forget it. *Labs 1 through 6 can be demonstrated sequentially without ever rewiring the board.*

Thus, time usually spent in troubleshooting a spaghetti forest of student wiring can be used more profitably, focussing on design principles and practices. This is not all to the good, of course. Hands-on wiring of devices teaches something about circuit construction that you don't get in software-based design entry. The DAC interface lab (Lab 7) gives some opportunity for circuit breadboarding, where students get some practice finding chip pin numbers, wiring discrete components, and connecting (or forgetting to connect) power supplies to circuits.

Software and Hardware Resources

The MAX+PLUS II software is bundled with these labs and can be copied freely. Each individual installed copy must be activated by an authorization code or license file available by e-mail from the Altera web site (**www.altera.com**). The Altera UP-1 circuit board is available from Altera's University Program for sale to students or can be requested by educational institutions either for purchase or on a donation basis for those institutions that are members of the Altera University Program. Institutions can also request donations of full-version software and CPLD chips.

MAX+PLUS II Design Files

Two sets of design and programming files are available:

1. **For Instructors:** a full set of required Graphic Design Files (gdf), VHDL files (vhd) and Programmer Object Files (pof) for all labs, the CPLD tutorial, and the design case study. The **gdf** and **vhd** files are uncompiled designs, without pin assignments. (Including pin assignments in uncompiled files creates some difficulties when transferring files to a computer other than the one on which it was created.) The **pof** files are compiled designs with full pin assignments that can be downloaded to the EPM7128S (MAX7000S family) CPLD on the Altera UP-1 board. These are available through the Delmar website at www.electronictech.com.

2. **For Students:** a set of design files for the CPLD tutorial and a limited set for some of the earlier labs in this series: **2bit_cmp.pof, debounce.vhd,** and **clockdiv.vhd.** These are available on the accompanying CD.

Programming CPLDs Using MAX+PLUS II

In order to take a digital design from the idea stage to the programmed silicon chip, we must go through a series of steps known as the CPLD Design Cycle. These include **design entry, simulation, compiling, fitting,** and **programming.** All steps require the use of PLD software, such as Altera's MAX+PLUS II, a **suite** of **software tools** used to perform the various tasks of the design cycle.

We will be using MAX+PLUS II as a vehicle for learning the concepts that relate to PLD design and programming. The **target devices** for our designs will be two Altera Complex Programmable Logic Devices (CPLDs), both installed on a circuit board available from Altera called the University Program Design Laboratory Package. We will generally refer to this board as the **Altera UP-1 Board.** (Any other board with an EPM7128S CLPD can also be used.)

In the remaining part of this tutorial, we will learn how to enter a design in MAX+PLUS II in both graphical and text format and how to compile the design and **download** it into one of the CPLDs on the Altera UP-1 circuit board.

Graphic Design File

One way of entering PLD designs is to create a **Graphic Design File.** This type of file contains a representation of a digital circuit, such as the "majority vote circuit" in Figure 1, showing components and their interconnections, as well as specifying the input and output names of the circuit.

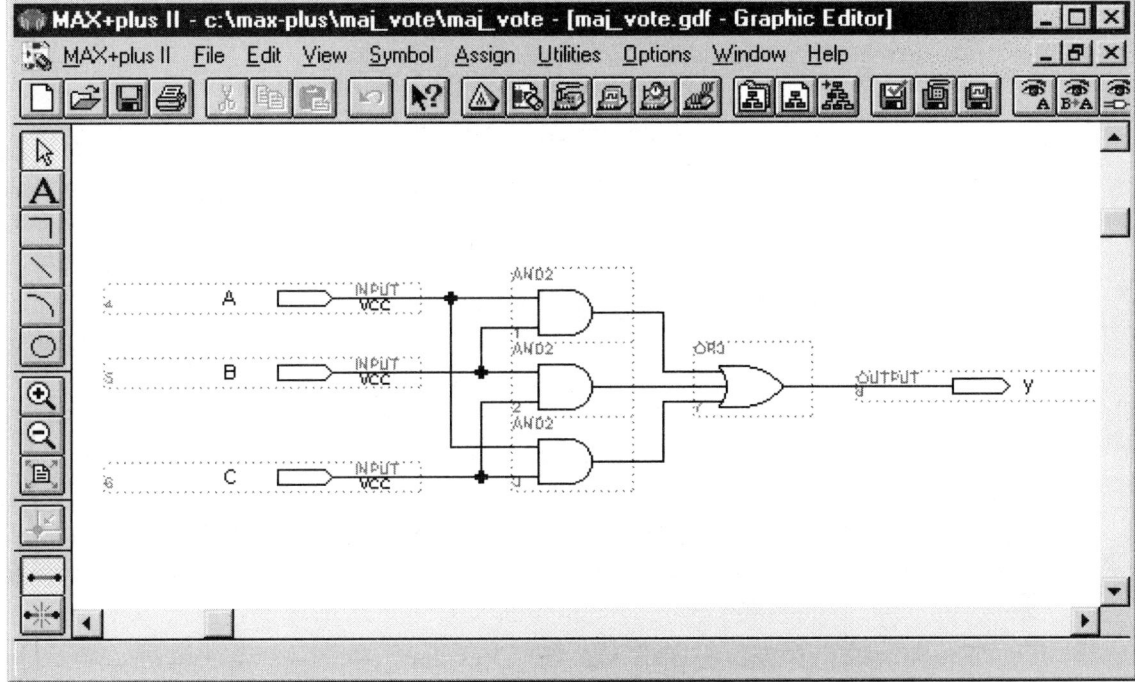

Figure 1 MAX+PLUS II Graphic Editor Screen Showing Majority Vote Circuit

MAX+PLUS II automatically generates a number of files to keep track of the PLD programming information represented by the Graphic Design File. These files, taken together, represent a **project** in MAX+PLUS II. All operations required to create a programming file for a CPLD are performed on a project, not a file. Thus, it is important during the design process to keep track of what the current project is.

When creating a new file, make it standard practice to first **Save** the file, then **Set Project to Current File**. (These functions are both found under the **File** menu of MAX+PLUS II.) If you form this habit, you (and MAX+PLUS II) will always know what the current project is. If you don't, you will find that you are saving or compiling some other project and wondering why your last set of changes didn't work.

Another good practice is to create a new Windows folder for each new design that you enter. Since MAX+PLUS II creates many files in the design process, the folders would become unmanageable if designs were not kept in separate folders.

In the following sections we will go through the process of creating a file in detail, using the majority vote circuit of Figure 1 as an example. This circuit produces a HIGH output when two out of three inputs are HIGH. The example assumes that MAX+PLUS II is properly installed on your computer and running. For installation instructions, see the instructions accompanying the *Student Edition of MAX+PLUS II* or the *MAX+PLUS II Installation* section of *MAX+PLUS II Getting Started,* available from Altera.

Entering Components

To create a Graphic Design File, click the **New File** icon on the tool bar or choose **New** on the MAX+PLUS II **File** menu. The dialog box, shown in Figure 2, appears. Select **Graphic Editor File** and choose **OK**.

Maximize the window and click the **Save** icon or choose **Save As** or **Save** from the

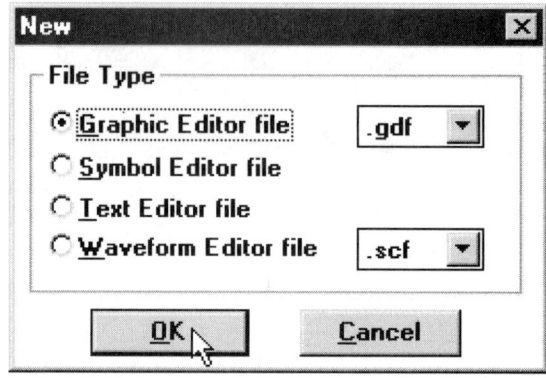

Figure 2   New File Dialog Box

File menu. In the dialog box shown in Figure 3, save the file in a new folder (e.g. *drive:\your_ folder*\maj_vote\maj_vote.gdf) and choose **OK**. (If you have not created the new folder, just type the complete path name in the **File Name** box. MAX+PLUS II will create a new folder.) Click the icon to **Set Project to Current File** or choose this action from the **File, Project** menu.

The first design step is to lay out and align the required components. We require three 2-input AND gates, a 3-input OR gate, three input pins, and one output pin. These basic

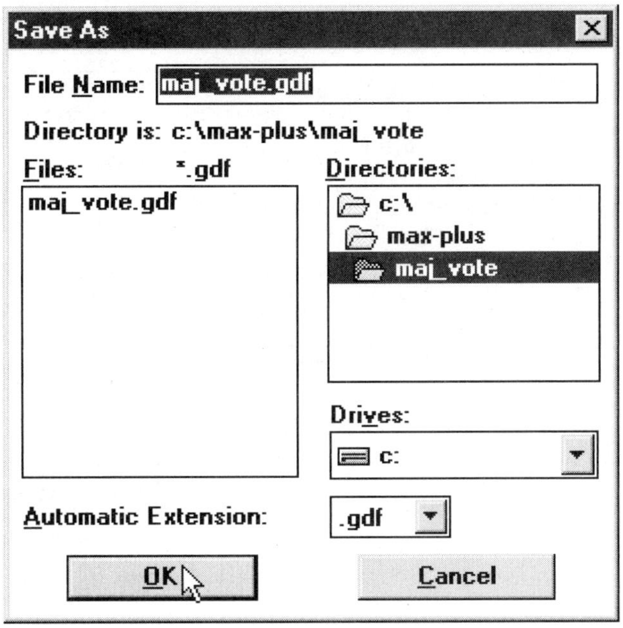

Figure 3   "Save As" Dialog Box

components are referred to as **primitives**. Let us start by entering three copies of the AND gate primitive, called **and2**.

Click the left mouse button to place the cursor (a flashing square) somewhere in the middle of the active window. Right-click to get a pop-up menu, shown in Figure 4, and choose **Enter Symbol**. The dialog box in Figure 5 appears. Type **and2** in the **Symbol Name** box and choose **OK**. A copy or **instance** of the and2 primitive appears in the active window.

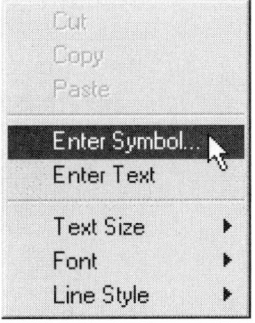

Figure 4   Pop-up Menu
for Entering a Symbol

You can repeat the above procedure to get two more instances of the and2 primitive, or you can use the **Copy** and **Paste** commands. These are the same icons and **File** commands as for other Windows programs. Highlight the and2 symbol by clicking it. Right-click the symbol to get the pop-up menu shown in Figure 6 and choose **Copy**. You can also click the **Copy** icon on the toolbar or use the **Copy** command in the **File** menu.

Paste an instance of the primitive by clicking to place the component, then right-clicking to bring up the menu shown in Figure 7. Choose **Paste**. The component will appear at the cursor location, marked in Figure 7 by the square at the top left corner of the pop-up menu.

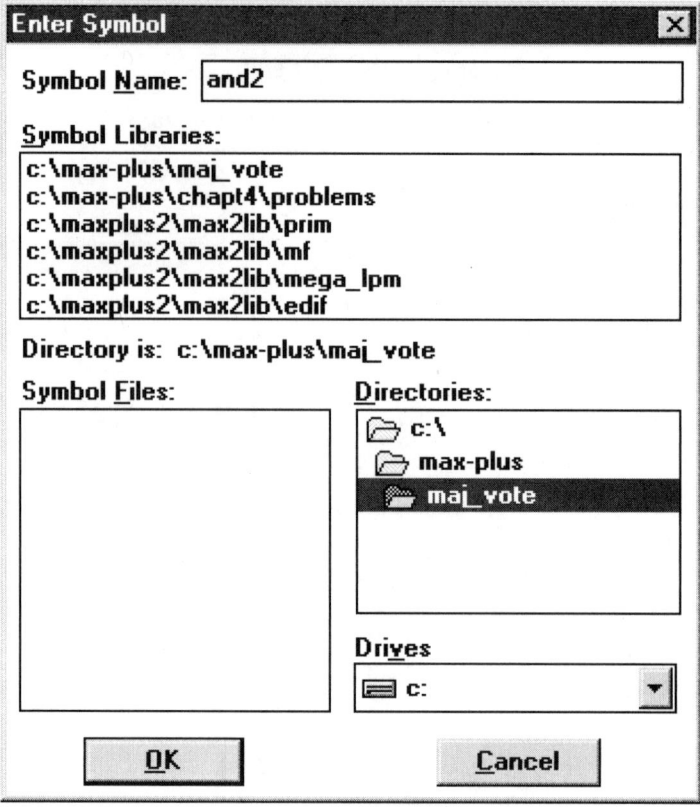

Figure 5   "Enter Symbol" Dialog Box

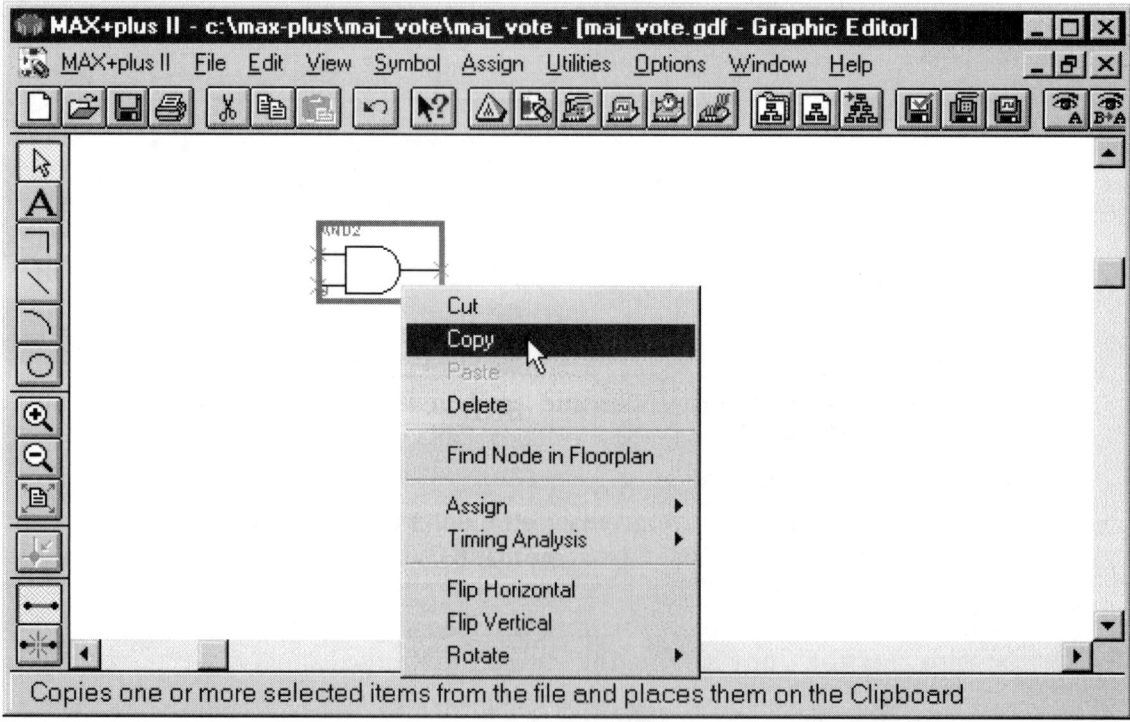

Figure 6   Copying a Symbol

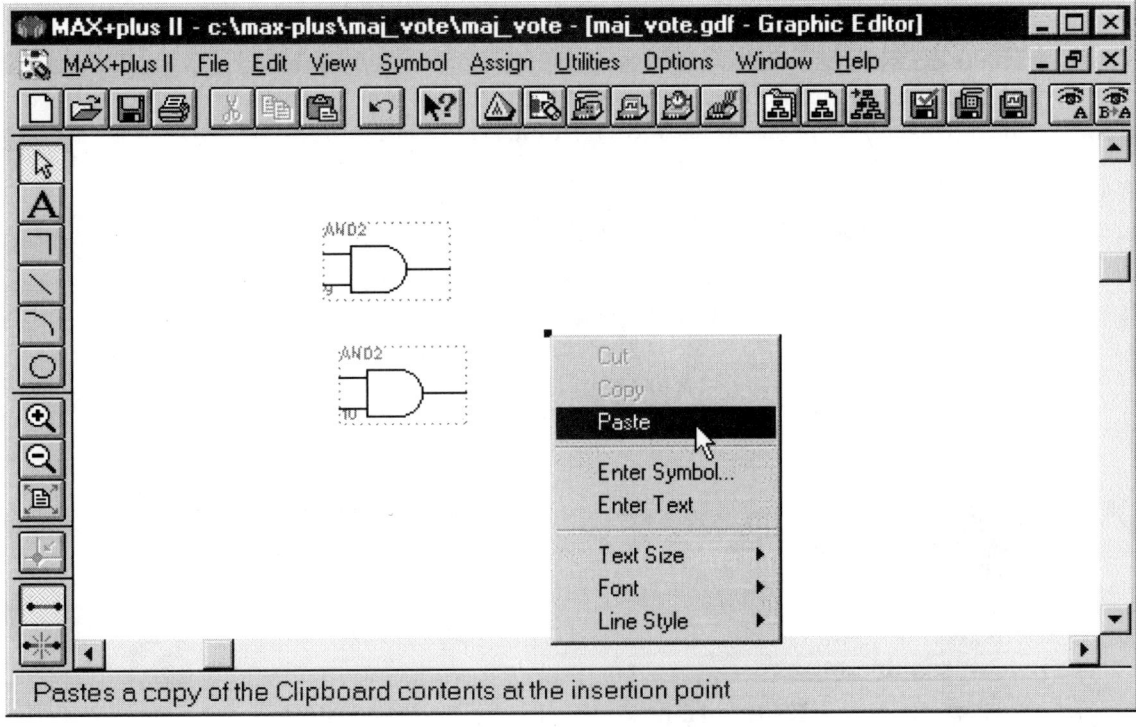

Figure 7   Pasting a Symbol

Enter the remaining components by following the **Enter Symbol** procedure outlined above. The primitives are called **or3, input,** and **output.** When all components are entered we can align them, as in Figure 8, by highlighting, then dragging each one to a desired location.

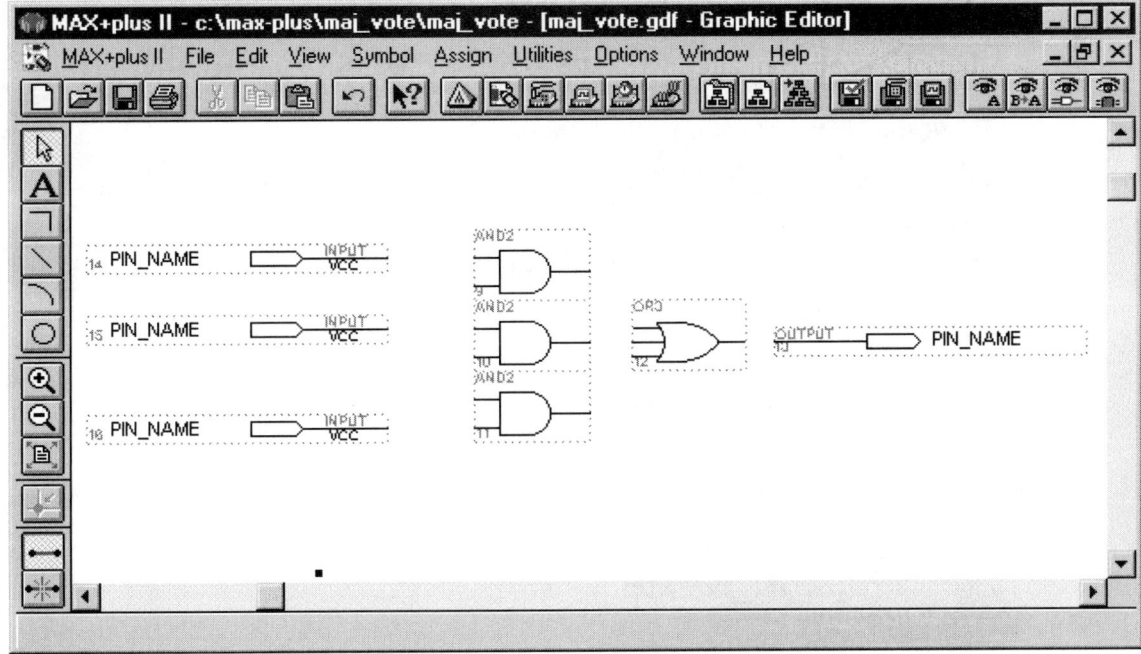

Figure 8   Aligning Components

Connecting Components

To connect components, click over one end of one component and drag a line to one end of a second component. When you drag the line, a horizontal and a vertical broken line mark the cursor position, as shown in Figure 9. These lines help you align connections properly.

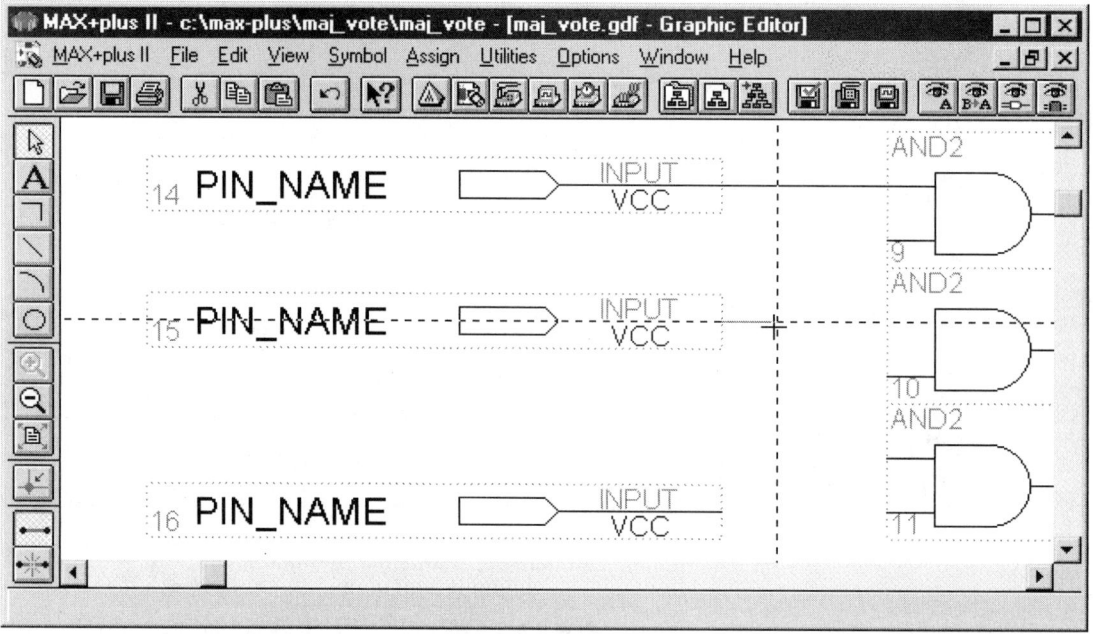

Figure 9   Connecting Components

A line will automatically make a connection to a perpendicular line, as shown in Figure 10. A line can have one 90-degree bend, as at the inputs of the AND gates. If a line requires two bends, such as shown at the AND outputs in Figure 11, you must draw two separate lines.

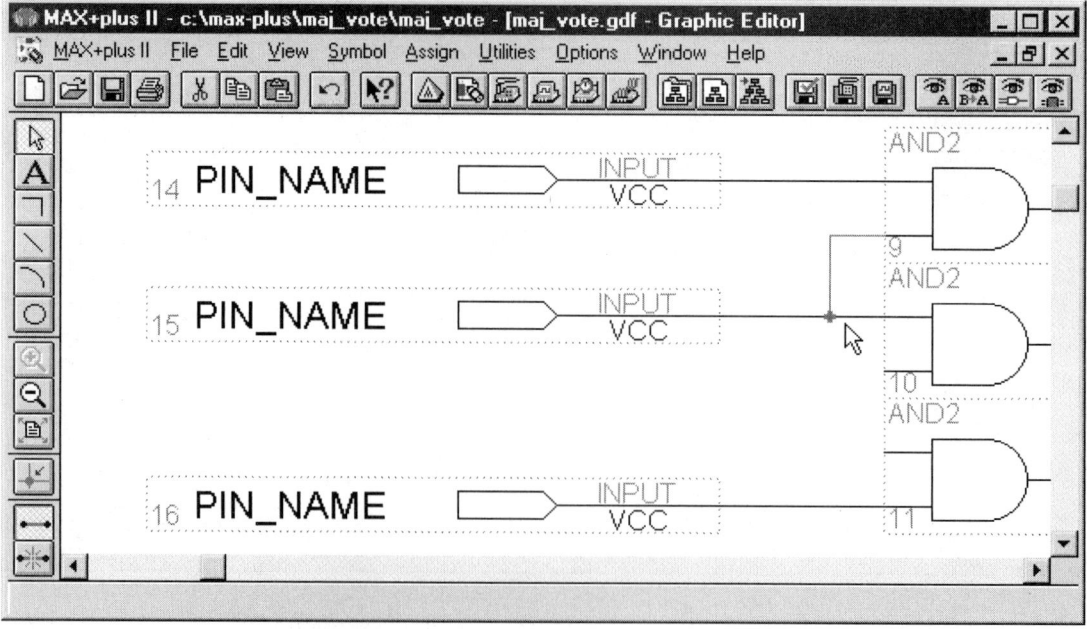

Figure 10   Perpendicular Lines

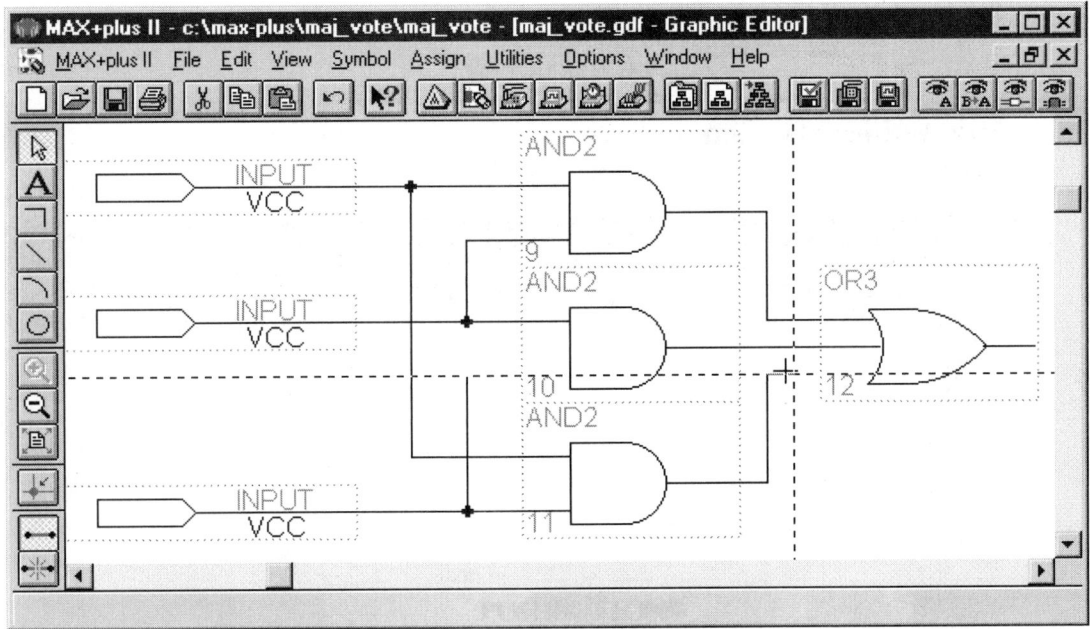

Figure 11   Lines with two 90-degree bends

Assigning Pin Names

Before a design can be compiled, its inputs and outputs must be assigned names. We could also specify pin numbers, if we wished to make the design conform to a particular CPLD, but it is not necessary to do so at this stage. It may not even be desirable to assign pin numbers, since the design we enter can be used as a component or subdesign of a larger circuit. We may also wish MAX+PLUS II to assign pins to make the best use of the CPLD's internal resources. At any rate, we will leave this step out for now.

Figure 12 shows the naming procedure. Pins A and B have already been assigned names. Highlight a pin by clicking on it. Right-click the highlighted pin and choose **Edit**

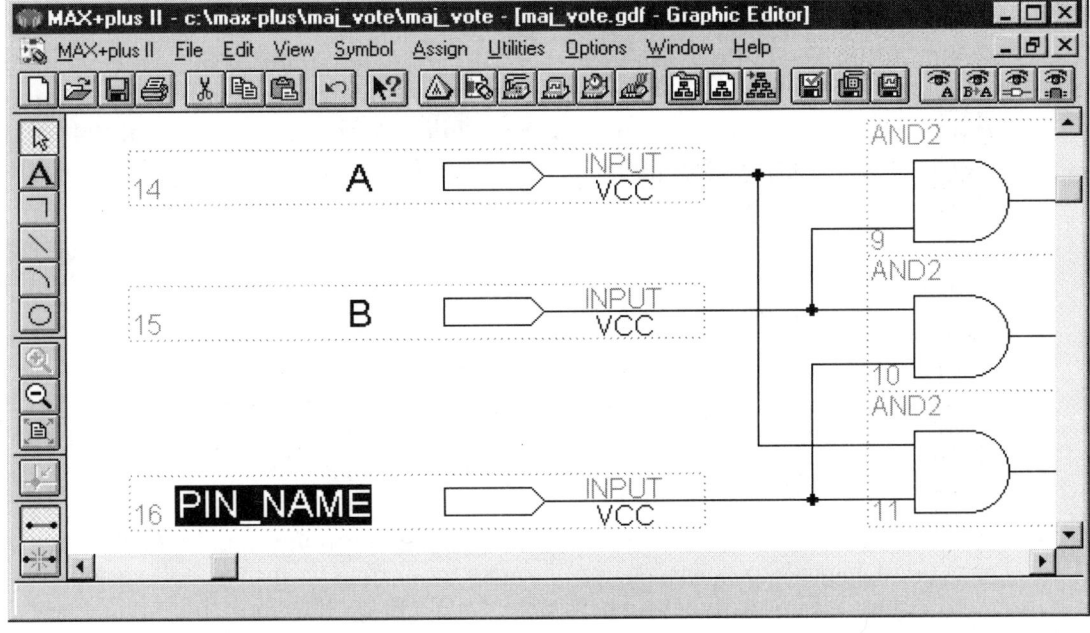

Figure 12   Assigning Pin Names

Pin Name from the pop-up menu. You could also double-click the pin name to highlight it. Type in the new name.

Hierarchical Design

A MAX+PLUS II Graphic Design File can be used as part of a **hierarchical design**. That is, it can be represented as a component in a higher-level design. Figure 13 shows a **gdf** that is constructed as a hierarchical design. It contains two majority vote circuits whose outputs are combined in an AND gate. Thus, the output would be HIGH if two out of three inputs were HIGH on *both* blocks labeled **maj_vote**. These blocks are complete designs in their own right, and thus form a lower level of the design hierarchy.

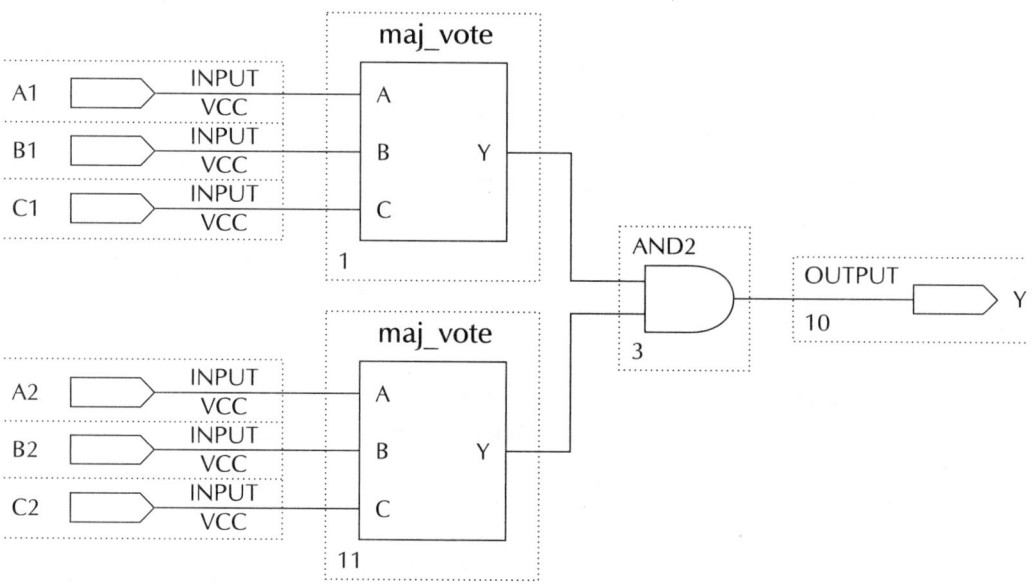

Figure 13   Hierarchical Desgin

Default Symbols and User Libraries

We can create a **default symbol** for the majority vote circuit of Figure 1 from the MAX+PLUS II **File** menu, as shown in Figure 14. This action will create a symbol file with the same name as the Graphic Design File and the extension **sym**. Before creating the symbol, make sure that the **gdf** is saved and that the project is set to the current file. The symbol can be embedded into a **gdf**, as in Figure 13.

Before we can use the new symbol, we must make sure that MAX+PLUS II knows where to find it. We can do this by setting a path to a **user library**, which is simply the Windows folder containing the **sym** file.

From the **Options** menu in MAX+PLUS II, select **User Libraries**. In the resultant dialog box, shown in Figure 15, select the appropriate drive and directories by double-clicking on the name in the **Directories** box. When the desired directory appears in the **Directory Name** box, click **Add**, then **OK**.

Note   If you are using MAX+PLUS II on a shared computer (e.g. in a computer lab), you might wish to delete all user libraries other than your own. A library that points to another user's directory can cause MAX+PLUS II to look there before (or instead of) looking in your directory, resulting in the apparent inability of MAX+PLUS II to find your file.

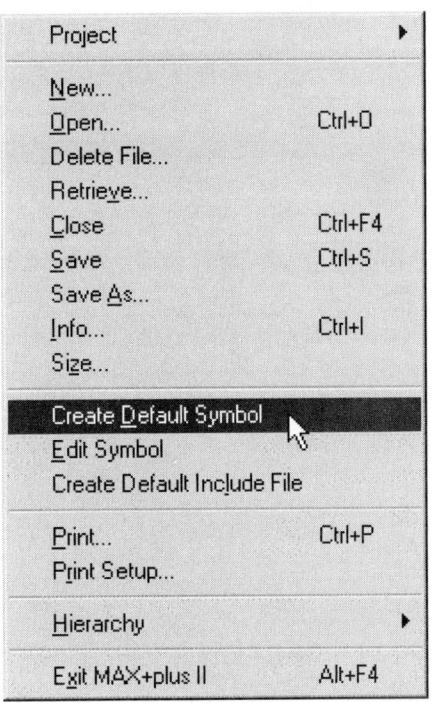

Figure 14   File Menu

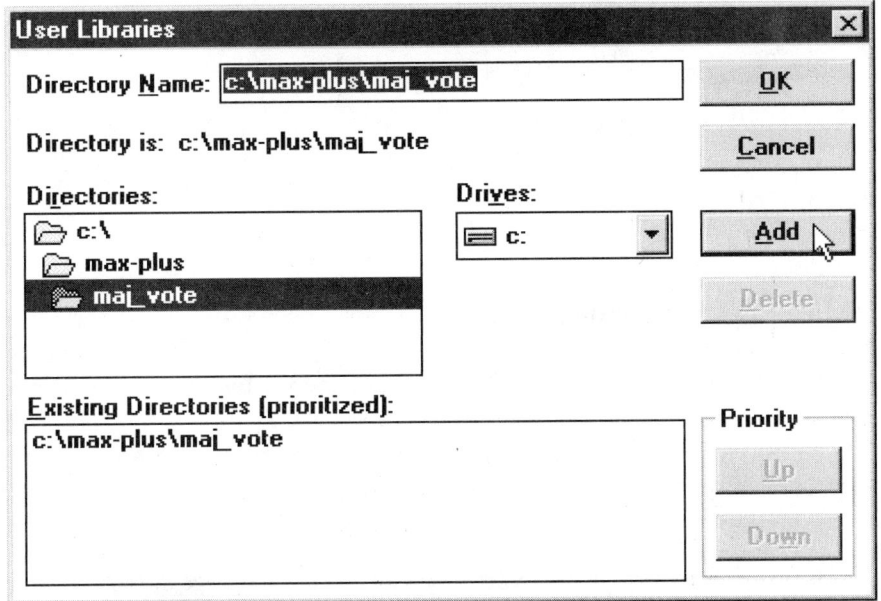

Figure 15   "User Libraries" Dialog Box

The circuit in Figure 13 is saved as **2votes.gdf**. If we double-click on either symbol labeled **maj_vote**, the MAX+PLUS II Graphic Editor will bring the file **maj_vote.gdf** to the foreground. Thus, we say that **2votes.gdf** is at the **top level** of the current hierarchy.

We can extend the hierarchy further by making a symbol for **2votes.gdf** and embedding it in a higher-level file called **4votes.gdf**, shown in Figure 16. This circuit generates a HIGH output if (two out of three of (A11, B11, C11) are HIGH AND two out of three of (A21, B21, C21) are HIGH) OR the same is true for (A12, B12, C12) AND (A22, B22, C22). If we double-click on either symbol for **2votes**, the Graphic Editor will bring the file **2votes.gdf** to the foreground.

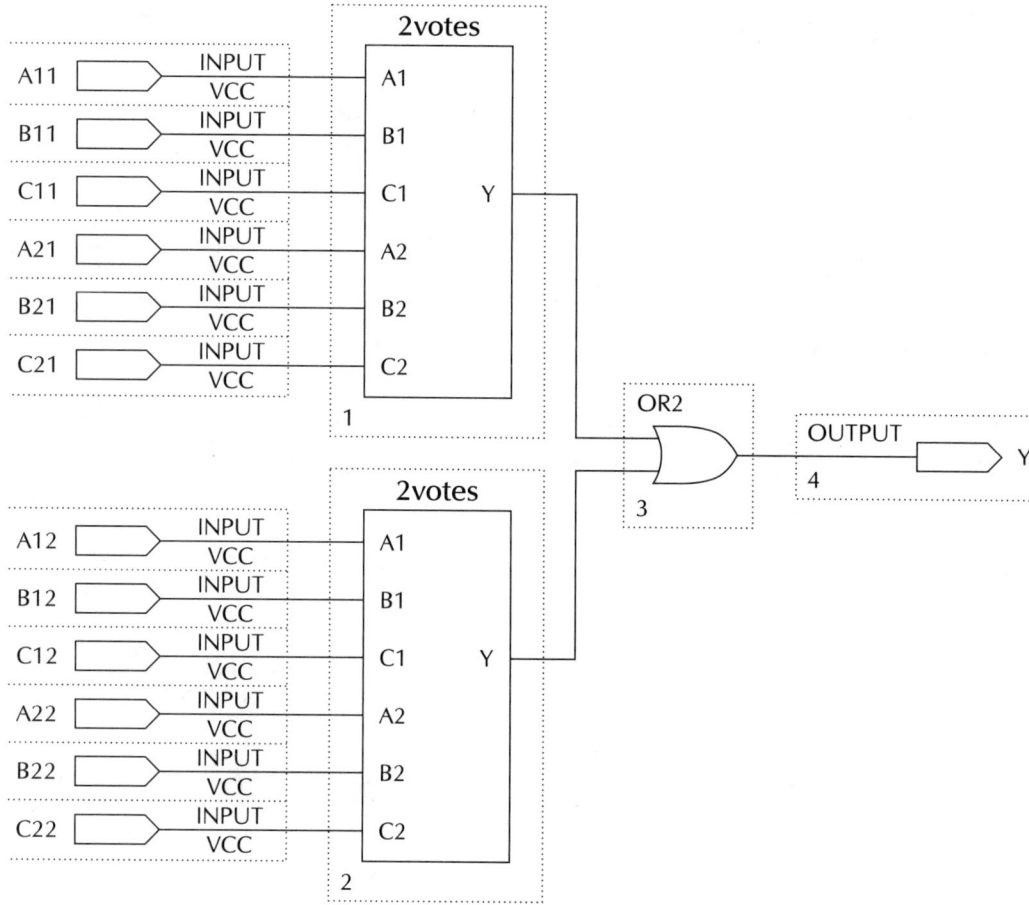

Figure 16   Further Level of Hierarchy

MAX+PLUS II can display the hierarchy of a design. To see the hierarchy structure, click the **Hierarchy** icon on the MAX+PLUS II toolbar (the yellow pyramid) or choose **Hierarchy Display** from the **MAX+PLUS II** menu. Figure 17 shows the hierarchy for the project **4votes**. Note that the highest level has two subdesigns, each of which breaks down further into two subdesigns. Thus, using hierarchical design and symbols for **gdf** or other design files allows us to create multiple instances of a basic design (**maj_vote.gdf**).

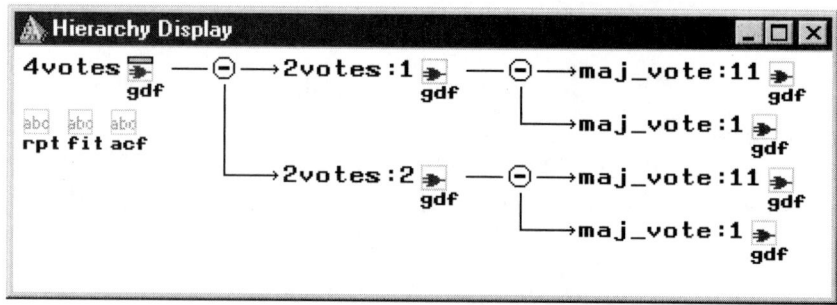

Figure 17   Hierarchy Display

Text Design File (VHDL)

An alternative, and ultimately more powerful, way to create PLD designs is through the use of a text-based design tool, such as Altera's **AHDL** or the industry-standard **VHDL**. These languages can be used to create hierarchical designs, either as components in graphic or text files or as higher-level files containing other designs.

AHDL, while very easy to use, has a much narrower application than VHDL because it is one of many proprietary tools on the market aimed at the programming requirements of a particular manufacturer's line of CPLDs. Since VHDL is an industry-standard language and the MAX+PLUS II compiler supports both languages, we will concentrate on VHDL.

VHDL was originally developed by defense contractors in the U.S. and is now the required standard for all **ASICs (Application Specific Integrated Circuits)** designed for the U.S. military. It has been standardized by the Institute of Electrical and Electronics Engineers (IEEE) and has been enjoying increasing popularity in the electronics design community. The original VHDL standard was written in 1987 and updated in 1993 (IEEE Std 1076-1993), with a new revision due soon.

Entity and Architecture

Every VHDL file requires at least two structures: an **entity** declaration and an **architecture** body. The entity declaration defines the *external* aspects of the VHDL function; that is, the input and output names and the name of the function. The architecture body defines the *internal* aspects; that is, how the inputs and outputs behave with respect to one another and with respect to other signals or functions that are internal only.

The VHDL file for a majority vote circuit is shown below. For more detail in creating this and other VHDL designs, refer to the document *Introduction to VHDL,* also part of this lab series.

```
-- maj_vot2.vhd
-- VHDL implementation of a majority vote circuit

-- Library contains standard VHDL models
LIBRARY ieee;
USE ieee.std_logic_1164.ALL;

 -- Entity defines inputs and outputs
ENTITY maj_vot2 IS
  PORT (
        a, b, c      : IN   STD_LOGIC;
        y            : OUT  STD_LOGIC);
END maj_vot2;

-- Architecture describes input/output relationship
ARCHITECTURE majority OF maj_vot2 IS
BEGIN
        y <= (a and b) or (b and c) or (a and c);
END majority;
```

Integrating VHDL and Graphical Design Components

We can create a default symbol for the VHDL majority vote function, much as we did for the same function in the Graphic Design File. In the Text Editor **File** menu, select **Create Default Symbol**. We can integrate this new symbol into a two-level majority vote circuit, as shown in Figure 18. This circuit contains primitives (AND gate, input pins, and output pin), a **gdf** symbol (maj_vote), and a symbol created from a VHDL file (MAJ_VOT2). Double-clicking on either symbol will bring forward its original design file.

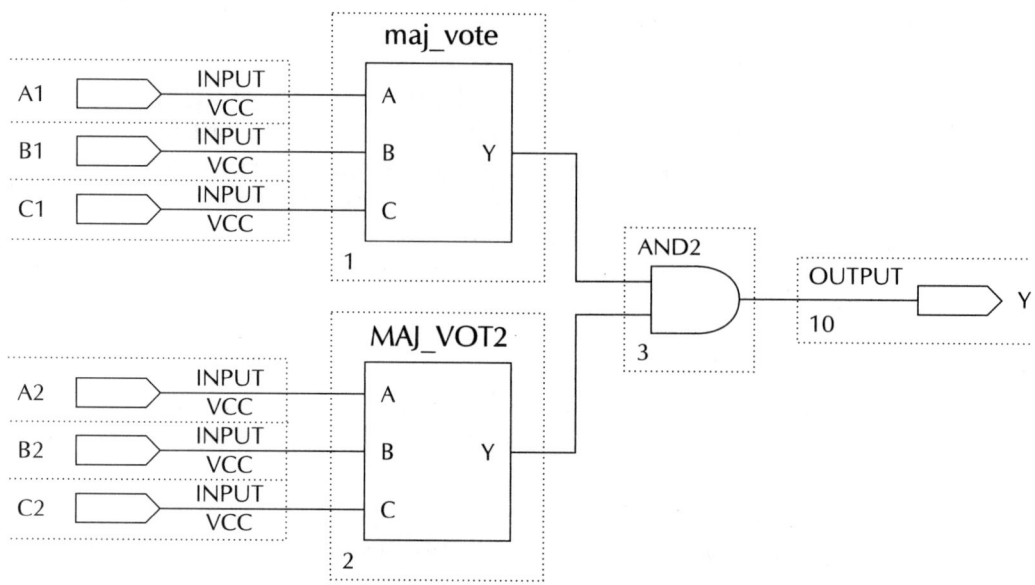

Figure 18   Hierarchical Design Incorporating Graphical and VHDL Components

Assigning Device and Pin Numbers

Before we can program our majority vote circuit into hardware, we must assign the input and output pin numbers and the device part number. Before proceeding with this step, make sure to save the file and set the project to the current file.

To assign a pin number, click on the pin to highlight it, then right-click to see the pop-up menu in Figure 19. Choose **Assign**, then **Pin/Location/Chip**. You can also do this from the **Assign** menu at the top of the screen.

We can assign device and pin numbers in the dialog box in Figure 20. Before MAX+PLUS II allows us to assign pin numbers, we must choose our device. The Student Edition of MAX+PLUS II only allows two choices: EPM7128SLC84-7 (MAX 7000S family) or EPF10K20RC240-4 (FLEX 10K series). Choose **Assign Device**. In the resultant dialog box, shown in Figure 21, select EPM7128SLC84-7 and click **OK**.

Type **A1** in the **Node Name** box in Figure 20, **12** in the **Pin** box and click **Add**. Type **B1** in the **Node Name** box, assign this name to pin16, and click **Add**. Repeat this procedure until all names are assigned, as in Table 1. When all assignments are complete, click **OK**.

Figure 22 shows the input pin assignments as they appear in the **gdf** file.

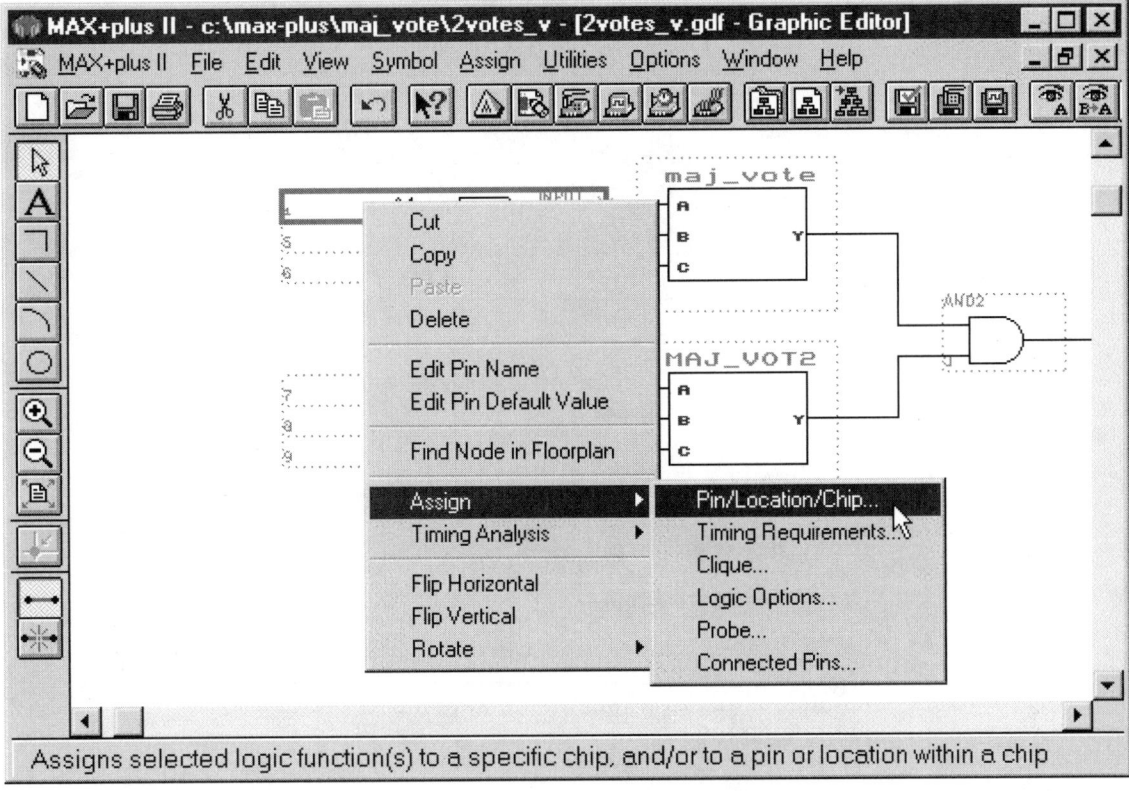

Figure 19   Pin Assignment Pop-Up Menu

Figure 20   Pin Assignment Dialog Box

Table 1
Pin Assignments for Majority Vote Circuit

PIN NAME	PIN NUMBER
A1	12
B1	16
C1	18
A2	15
B2	17
C2	21
Y	4

Figure 21   Assign Device Dialog

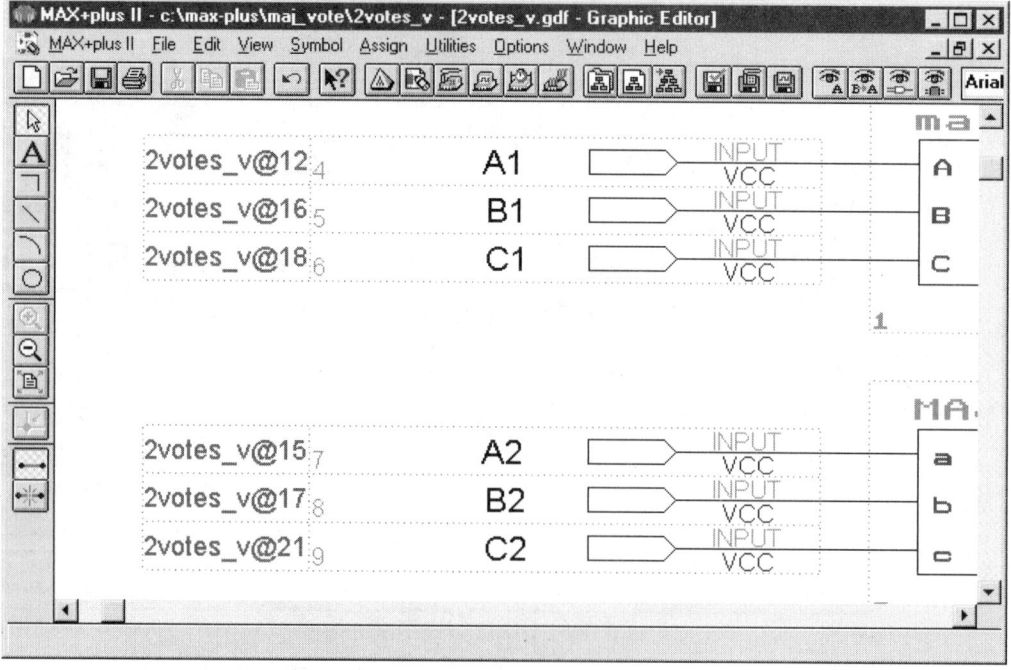

Figure 22   Pin Numbers in a gdf File

Compiling MAX+PLUS II Files

The MAX+PLUS II compiler converts design entry information into binary files that can be used to program a CPLD. The compiler has a number of settings that can be chosen prior to the actual compile process. Figure 23 shows some of the settings that should be selected from the **Processing** menu of the **Compiler** window. You can open the **Compiler** window from the **MAX+PLUS II** menu or by clicking the **Compiler** button on the toolbar at the top of the screen.

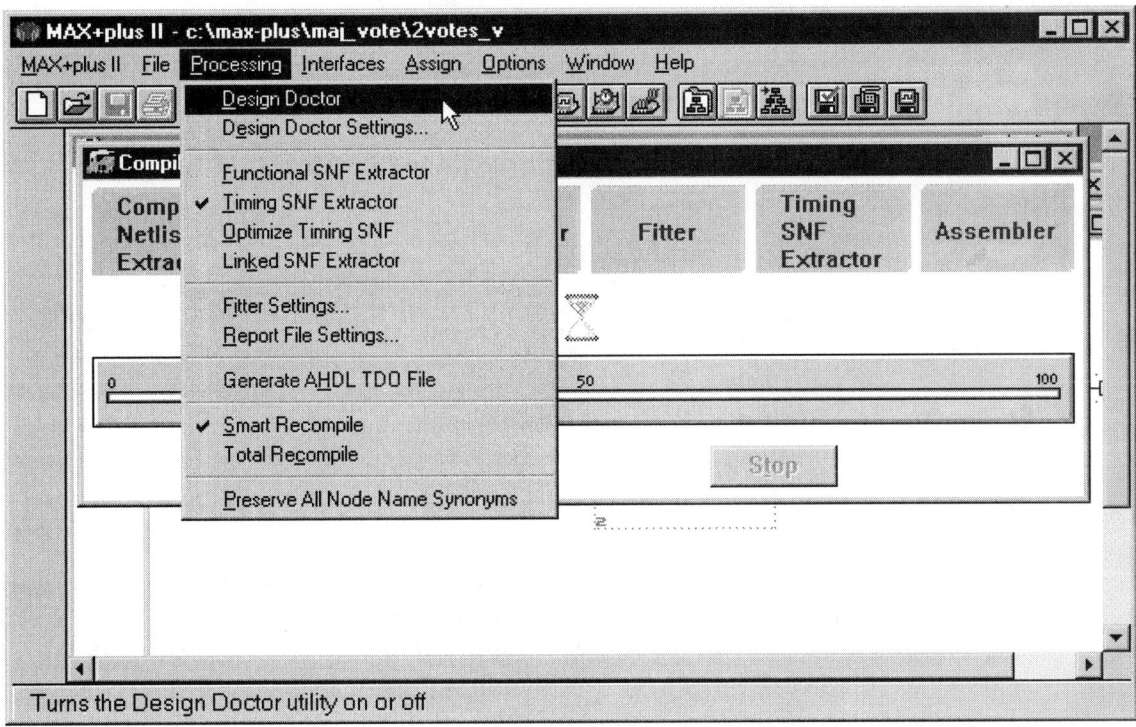

Figure 23   Setting Compiler Options

Design Doctor is a utility that checks for adherence to good design practice and will warn you of any bad design choices. (Design Doctor will not stop the design from compiling, but will suggest potential problems that could result from a particular design.) The **Timing SNF Extractor** creates a Simulation Netlist File, which is required to perform a timing simulation of the design. We will perform this step in later MAX+PLUS II designs. (If you are not able to select the **Timing SNF Extractor,** then uncheck the **Functional SNF Extractor** option.) **Smart Recompile** allows the compiler to use previously compiled portions of the design to which no changes have been made. This allows the compiler to avoid having to compile the entire design each time a change is made to one part of the design.

To start the compile process, click **Start** in the **Compiler** window. While in progress, the window will look something like Figure 24. Messages of three types may appear during the compile process. **Info** messages (green type) are for information only. **Warning** messages (blue type) tell you of potential, but nonfatal, problems with the design. **Error** messages (red type) inform you of design flaws that render the design unusable. A CPLD can still be programmed if the compiler generates **info** or **warning** messages, but not if it generates an **error.**

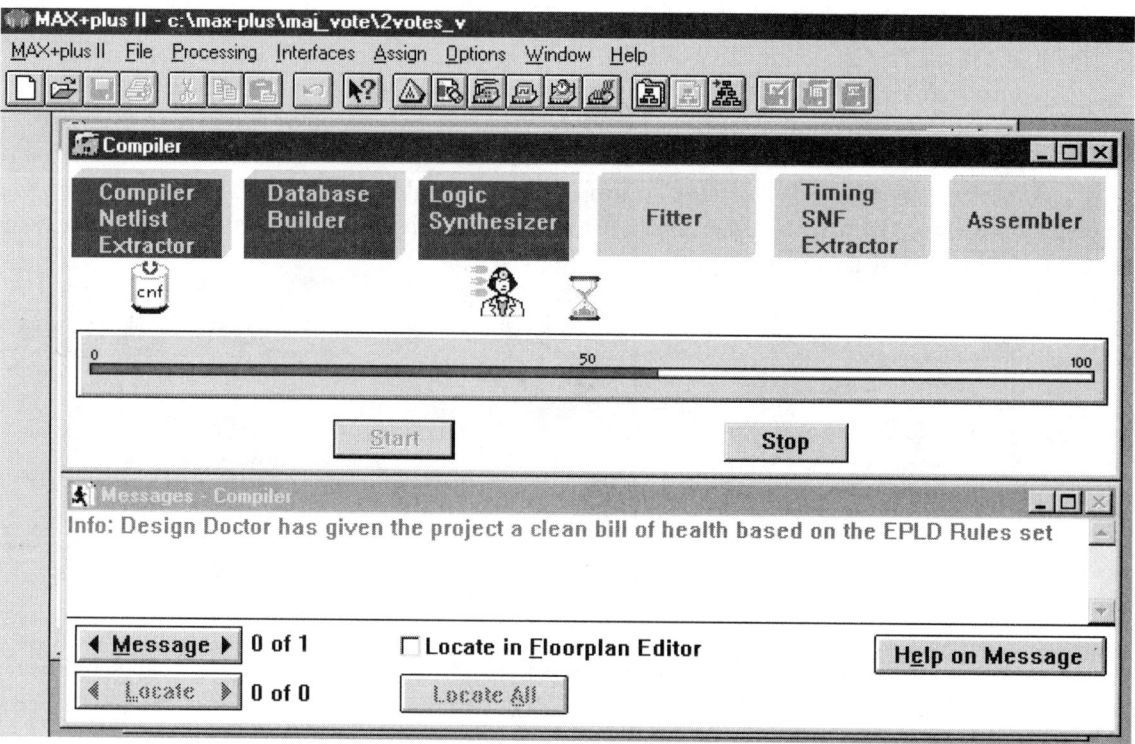

Figure 24   Compiler Window

Depending on the device chosen, the compiler generates either a **Programmer Object File (pof)** or **SRAM Object File (sof)**. The **pof** is used to *program* a MAX-series PLD. The **sof** is used to *configure* a FLEX-series PLD. The difference is that the MAX device is **nonvolatile**, that is, it retains its programming information after the power has been removed. The FLEX-series device is **volatile**, meaning that its programming information must be loaded each time the device powers up.

Programming CPLDs on the Altera Up-1 Circuit Board

The CPLDs on the Altera UP-1 circuit board are programmed via the programming software in MAX+PLUS II and a ribbon cable called the **ByteBlaster**. The ByteBlaster connects the parallel port of the PC running MAX+PLUS II to a 10-pin male socket that complies with the **JTAG** standard. This standard specifies a four-wire interface, originally developed for testing chips without removing them from a circuit board, but can also be used to program or configure PLDs.

PLDs that can be programmed or configured while installed on a circuit board are called **In-System Programmable (ISP)** or **In-Circuit Reconfigurable (ICR)**. ISP is used to refer to nonvolatile devices, such as MAX7000S; ICR refers to volatile devices, such as FLEX10K.

The JTAG interface has four wires, as well as power and ground connections. Data are sent to a device from a JTAG controller (i.e. the PC) via the **TDI (Test Data In)** line. Data return from the device via **TDO (Test Data Out)**. The data transfer is controlled by **TMS (Test Mode Select)**. The process is driven from one step to the next by **TCK (Test Clock)**.

Multiple devices can be programmed in a **JTAG Chain**, whereby data can be sent through a chain of devices, the TDO of one device feeding the TDI of the next. This connection allows both CPLDs on the Altera UP-1 Board to be programmed at the same time. The UP-1 board also has a female 10-pin socket labeled JTAG out, which allows two or more boards to be chained together. The choice of programming one or more CPLDs, or the CPLDs on one or more UP-1 boards, is determined by the placement of four on-board jumpers. These jumper positions are explained in the *Altera University Program Design Laboratory Package User Guide.*

The operation of the JTAG port is controlled automatically by MAX+PLUS II, so further details are not necessary at this time. For further information on the JTAG interface, refer to *Altera Application Note 39, JTAG Boundary-Scan Testing in Altera Devices.*

MAX+PLUS II Programmer

To program a device on the Altera UP-1 board, set the jumpers to program the EPM7128S or configure the EPF10K20, as described in the UP-1 User's Guide. Connect the ByteBlaster cable from the parallel port of the PC running MAX+PLUS II to the 10-pin JTAG header. (You may have to run a 25-wire cable (male-D-connector-to-female-D-connector) to make it reach.) Plug an AC adapter (9-volt dc output) into the power jack of the UP-1 board.

Open the top-level file of the project you wish to download to the UP-1 board (e.g., **maj_vote.gdf**). Set the project to the current file. Invoke the MAX+PLUS II Programmer from the **MAX+PLUS II** menu or click the Programmer button (the icon showing the blue ribbon cable) on the MAX+PLUS II toolbar.

If you have never programmed a device with your copy of MAX+PLUS II, you will need to set up the hardware configuration. Click **Hardware Setup** in the **Options** menu to get the dialog box in Figure 25.

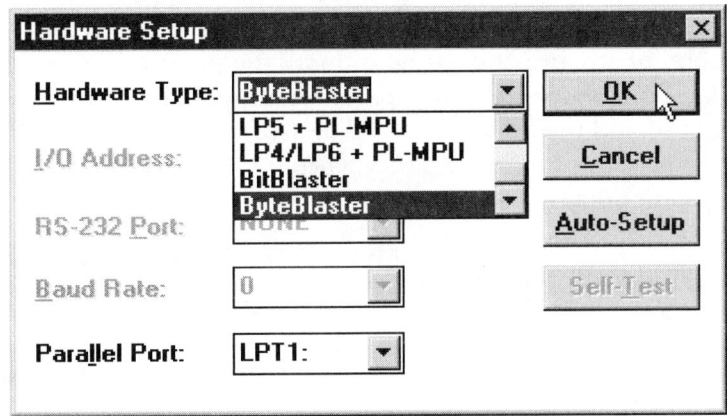

Figure 25   Hardware Setup Dialog

Select **ByteBlaster** in the **Hardware Type** box. Ensure that **Parallel Port** is the same as the port the ByteBlaster is plugged into (usually LPT1:). Click **OK.** (If you have a choice, configure your parallel port as an Enhanced Communications Port (ECP). For most users this step is not necessary, as the port is already configured this way.)

If the current project was compiled with the MAX7000S device selected, the **pof** file for the project will automatically be available. The programmer window will appear as in Figure 26. To download, click **Program** (MAX7000S).

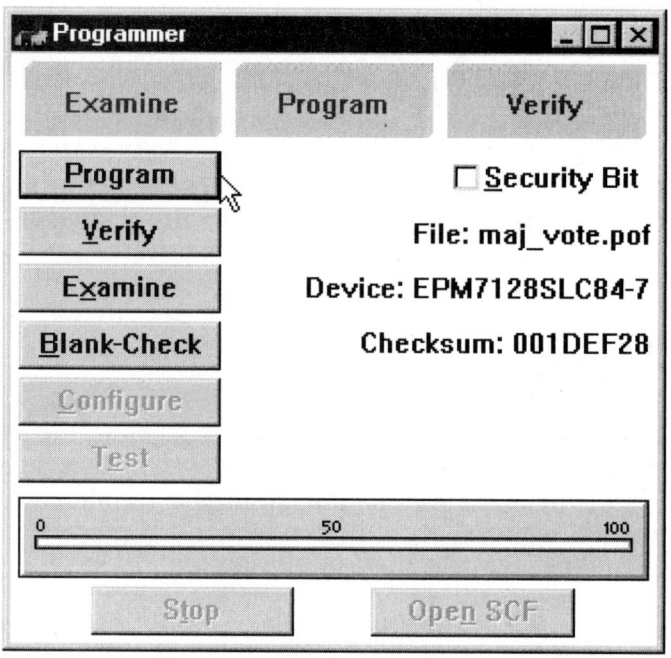

Figure 26   Programmer Dialog

If the project was compiled for the FLEX10K device, it must be configured via the **Multi-Device JTAG Chain** available in the **JTAG** menu. Select the **JTAG** menu, and choose **Multi-Device JTAG Chain**.

In the **Multi-Device JTAG Chain Setup** window, shown in Figure 27, select the pull-down menu for the device name. Select **EPF10K20RC240**. Choose **Delete All** to clear the box of any previous programming file names. Choose the **Select Programming File** button. Find and select the file *drive:*\max-plus\maj_vote\maj_vote.sof. Choose the **Add** button to add the SRAM Object File (**sof**) to the list. Choose the **Detect JTAG Chain Info** button to set up the hardware for programming. Choose **OK**. This will return you to the Programmer dialog box.

Make sure that the jumpers on the UP-1 board are set to configure the EPF10K20 device. Click the **Configure** button to download the binary information to the FLEX10K CPLD on the UP-1 Board.

At this point the design can be physically tested. If it was loaded to the MAX7000S chip, we can wire three DIP switches to the input pins, via the MAX Prototyping Headers. Pin numbers for these headers are listed in the UP-1 User's Guide. The output can be wired to an LED, keeping in mind that the Altera UP-1 LEDs are LOW-sense. That is, an LED illuminates with a logic LOW.

If the FLEX10K chip was configured, input DIP switches and output LEDs are hardwired to pins, as indicated in the UP-1 User's Guide. Pin assignments will have to correspond to these hardwired connections.

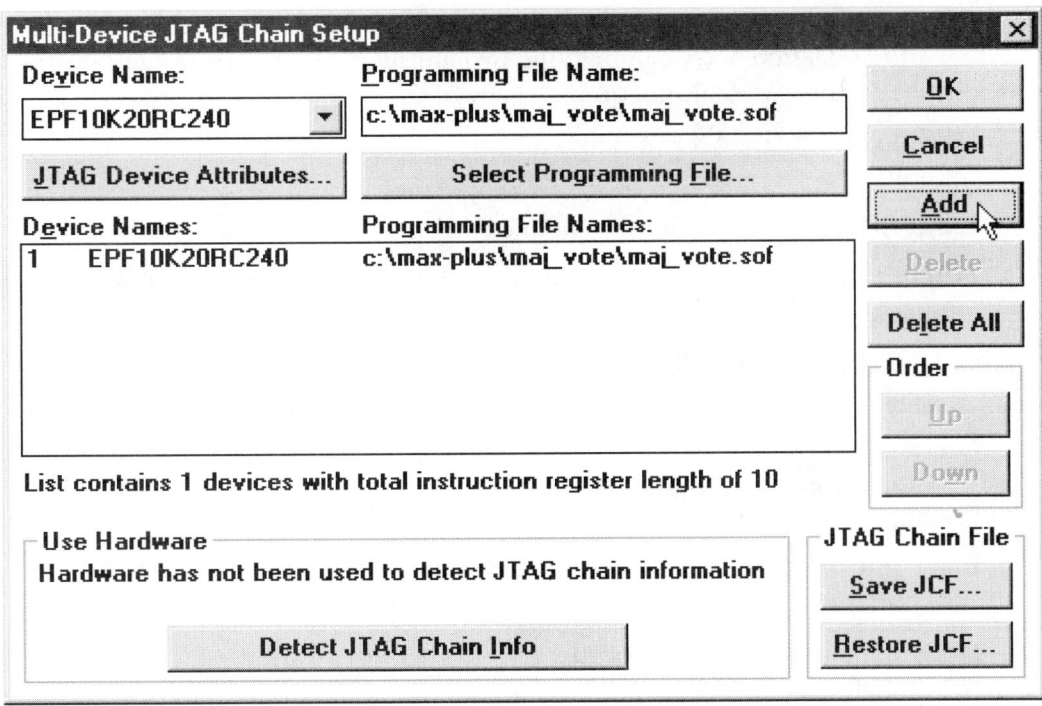

Figure 27   Multi-Device JTAG Chain Setup

Key Terms

AHDL (Altera Hardware Description Language)   Altera's proprietary text-entry design tool for PLDs.

Altera UP-1 Board   A circuit board, part of Altera's University Program Design Laboratory Package, containing two CPLDs and a number of input and output devices.

Architecture   A VHDL structure that defines the relationship between input, output, and internal signals or variables in a design.

ASICs (Application Specific Integrated Circuits)   Integrated circuits that are constructed for a specific design purpose. The term could refer to a PLD, although it usually means a custom-designed fixed-function device.

ByteBlaster   An Altera ribbon cable and connector used to program or configure Altera PLDs via the parallel port (LPT port) of an IBM PC or compatible.

Comment   Explanatory text in a VHDL (or other computer language) file that is ignored by the computer at compile time.

Compile   The process used by CPLD design software to interpret design information (such as a drawing or text file) and create required programming information for a CPLD.

Default Symbol   A graphical symbol that represents a PLD design as a block, showing only the design's inputs and outputs. The symbol can be used as a component in any Graphic Design File.

Design Entry   The process of using software tools to describe the design requirements of a PLD. Design entry can be done by entering a schematic or a text file that describes the required digital function.

Download   Program a PLD from a computer running PLD design and programming software.

Entity   A VHDL structure that defines the inputs and outputs of a design.

Fitting   Assigning internal PLD circuitry, as well as input and output pins, for a PLD design.

Graphic Design File (gdf)   A PLD design file in which the digital design is entered as a schematic.

Hierarchical Design   A PLD design that is ordered in layers or levels. The highest level of design contains components that are themselves complete designs. These components may, in turn, have lower-level designs embedded within them.

ICR   In-Circuit Reconfigurability. The ability of a PLD (such as a FLEX10K) to be configured without removing it from a circuit board.

IEEE Standard 1164   The standard which defines a variety of VHDL types and operations, including the STD_LOGIC and STD_LOGIC_VECTOR types.

Instance   A single copy of a component in a PLD design file.

ISP   In-System Programmability. The ability of a PLD (such as a MAX7000S) to be programmed without removing it from a circuit board.

JTAG   Joint Test Action Group. A standard that specifies the format for testing devices while they are installed in a system. The same standard also allows PLDs to be programmed or configured in a circuit via a four-wire interface.

JTAG Chain   Multiple JTAG-compliant devices whose TDI and TDO ports form a continuous chain connection. Such a chain allows multi-device programming.

Nonvolatile   Able to retain stored information after power is removed.

Primitives   Basic functional blocks, such as logic gates, used in PLD design files.

Programmer Object File (pof)   Binary file used to program a nonvolatile PLD, such as those of the Altera MAX series.

Programming   Transferring design information from the computer running PLD design software to the actual PLD chip.

Project   A set of MAX+PLUS II files associated with a particular PLD design.

Simulation   Testing design function by specifying a set of inputs and observing the resultant outputs. Simulation is generally shown as a series of input and output waveforms.

Software Tools   Specialized computer programs used to perform specific functions such as design entry, compiling, fitting, and so on. (Sometimes just called "tools".)

SRAM Object File (sof)   Binary file used to configure a volatile PLD, such as those of the Altera FLEX series.

Suite (of software tools)   A related collection of tools for performing specific tasks. MAX+PLUS II is a suite of tools for designing and programming digital functions in a PLD.

Syntax   The "grammar" of a computer language (i.e., the rules of construction of language statements).

Target Device   The specific PLD for which a digital design is intended.

TCK   Test Clock. The JTAG signal that drives the JTAG downloading process from one state to the next.

TDI   Test Data In. In a JTAG port, the serial input data to a device.

TDO   Test Data Out. The serial output data from a device in a JTAG port.

TMS   Test Mode Select. The JTAG signal that controls the downloading of test or programming data.

Top Level (of a Hierarchy)   The file in a hierarchy that contains components specified in other design files and is not itself a component of a higher-level file.

User Library   A folder containing symbols that can be used in a **gdf** file.

VHDL (VHSIC Hardware Description Language)   An industry-standard computer language used to model digital circuits and produce programming data for PLDs.

VHSIC   Very High Speed Integrated Circuit.

Volatile   A device is volatile if it does not retain its stored information after the power to the device is removed.

Introduction to VHDL

An alternative to schematic entry of PLD designs is to use a text-based design tool, such as the industry-standard **VHDL**. (VHDL stands for VHSIC Hardware Description Language. VHSIC = Very High Speed Integrated Circuit.) In VHDL, a designer creates a text file, framed within a certain set of rules known as the **syntax** of the language, and uses a compiler to create programming data. A VHDL design entity can be used in hierarchical designs, either as a component in graphic or text files or as a higher level file containing other design entities.

It is impossible to cover all aspects of VHDL in a short tutorial such as this. Rather than make a comprehensive survey of the language, we will look at the bare-bones essentials of the language plus a few useful constructs. We will also look at some practical circuit implementations using VHDL.

Entity and Architecture

Every VHDL file requires at least two structures: an **entity** declaration and an **architecture** body. The entity declaration defines the *external* aspects of the VHDL function; that is, the input and output names and the name of the function. The architecture body defines the *internal* aspects; that is, how the inputs and outputs behave with respect to one another and with respect to other signals or functions that are internal only.

VHDL Templates in MAX+PLUS II

MAX+PLUS II offers a shortcut to creating VHDL code in the form of a **Template Menu**. To choose a template, select the one desired from the **VHDL Template** dialog box, shown in Figure 28.

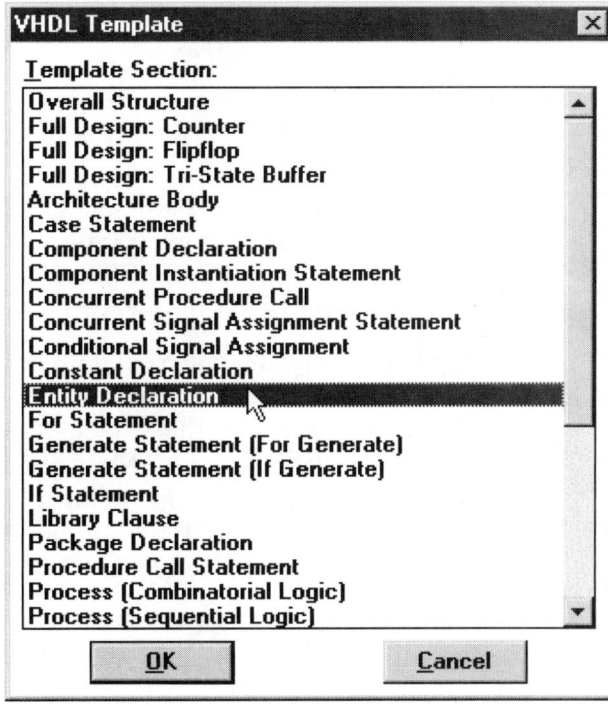

Figure 28   VHDL Template Dialog Box

Entity Declaration

Choosing the Entity Declaration template results in the following text:

```
ENTITY __entity_name IS
    PORT (
         __input_name, __input_name   : IN STD_LOGIC;
         __input_vector_name: IN STD_LOGIC_VECTOR(__high downto __low);
         __bidir_name, __bidir_name   : INOUT  STD_LOGIC;
         __output_name, __output_name : OUT    STD_LOGIC);
END __entity_name;
```

To convert this into a valid entity for our use, we delete the lines we do not need and substitute input and output names into the template. For example, let us encode a **majority vote circuit,** a combinational circuit whose output goes HIGH if two out of three inputs is HIGH. The Boolean equation for this circuit is:

$$Y = AB + BC + AC$$

For our majority vote circuit, we can modify the template to yield the entity declaration:

```
ENTITY maj_vote IS
    PORT (
        a, b, c : IN   STD_LOGIC;
        y       : OUT  STD_LOGIC);
END maj_vote;
```

The name of the entity, **maj_vote,** is given in the first and last lines of the entity declaration. The VHDL file that contains this entity should be named **maj_vote.vhd.** (MAX+PLUS II requires that a VHDL file be given the same name as an entity in the file.)

Ports, Type, and Libraries

In between the first and last statements of the entry declaration is a **port** definition, indicating which names are for input and which for output and the **type** assigned to each one. In this case the type is **STD_LOGIC** (Standard Logic). As indicated in the template, we can also have port names of type **STD_LOGIC_VECTOR,** which allow us to use a group of related inputs as a single multibit variable or bus. In addition to input and output ports, we can designate a port as **INOUT** (input or output), which can be used as a bidirectional port, or **BUFFER,** which is like an output but allows the output signal to be fed back into the circuit for use elsewhere.

The STD_LOGIC and STD_LOGIC_VECTOR types define the usual binary values of '0' and '1', as well as seven other logic states, including 'Z' (high impedance) and 'X' (unknown). This makes the STD_LOGIC types more widely useful than the BIT and BIT_VECTOR types which only include '0' and '1'. However, the BIT and BIT_VECTOR types can be used without including a **library** (that is, a collection of compiled VHDL design units) in the VHDL file. A number of other types are available in VHDL, including INTEGER, BOOLEAN, and CHARACTER.

The STD_LOGIC type is defined in a library called **ieee.** Since the STD_LOGIC type is defined by IEEE Standard 1164, there is a **package** of functions in the **ieee** library called **std_logic_1164.** The syntax for using this package is:

```
LIBRARY ieee;
USE ieee.std_logic_1164.ALL;
```

Architecture Body

The VHDL template for an architecture body yields the following text:

```
ARCHITECTURE a OF __entity_name IS
      SIGNAL __signal_name : STD_LOGIC;
      SIGNAL __signal_name : STD_LOGIC;
BEGIN
   -- Process Statement
   -- Concurrent Procedure Call
   -- Concurrent Signal Assignment
   -- Conditional Signal Assignment
   -- Selected Signal Assignment
   -- Component Instantiation Statement
   -- Generate Statement
END a;
```

The architecture body consists of an architecture name and the name of a previously declared entity. Internal **signals** and **variables** are declared in the architecture body, followed by a large variety of possible operations and functions. A signal is like a wire connecting portions of the VHDL design. It can be used globally throughout the design and is declared after the ARCHITECTURE clause, but before BEGIN. A variable is simply a name for a piece of working memory. It is local only to a specific portion of the VHDL file. A variable is declared inside the architecture body (after BEGIN).

As an introduction to signal assignments, let us look at the **concurrent signal assignment** statement. Concurrent means "simultaneous." The implication is that any number of concurrent signal assignments can be listed in a VHDL architecture body and the order in which they are evaluated does not depend on the order in which they are written, since all statements are concurrent. This is in contrast to **sequential statements**, in which order is important.

The Boolean equation for a 3-input majority vote circuit is $Y = AB + BC + AC$. In a concurrent signal assignment, we can write this operation as:

```
y <= (a and b) or (b and c) or (a and c);
```

The operator ($<=$) assigns the value of the right hand side of the equation to the left hand side. Whenever there is a change in a, b, or c, the statement is reevaluated and the new value is assigned to y. Note that VHDL logical operators (such as **and** and **or**) have equal precedence, so we must make the order of precedence explicit with parentheses.

The complete architecture body for the majority vote function, which we have named `majority`, is:

```
ARCHITECTURE majority OF maj_vote IS
BEGIN
y <= (a and b) or (b and c) or (a and c);
END majority;
```

The complete VHDL file for the majority vote circuit is shown below. The double dashes before the first two lines are to indicate that these lines are **comments**. There are also a few other comments to illustrate the use of VHDL

```
-- maj_vote.vhd
-- VHDL implementation of a majority vote circuit

-- Library contains type definitions for "standard logic"
LIBRARY ieee;
USE ieee.std_logic_1164.ALL;

-- Entity defines inputs and outputs
ENTITY maj_vote IS
PORT (
    a, b, c : IN   STD_LOGIC;
    y       : OUT  STD_LOGIC);
END maj_vote;

-- Architecture describes input/output relationship
ARCHITECTURE majority OF maj_vote IS
BEGIN
     y <=  (a and b) or (b and c) or (a and c);
END majority;
```

Another VHDL Example: Vector Notation

Figure 29 shows the logic diagram of a 2-line-to-4-line decoder. The circuit detects the presence of a particular binary code and makes one and only one output HIGH, depending on the value of the two-bit number D_1D_0. Write a VHDL file that describes the decoder.

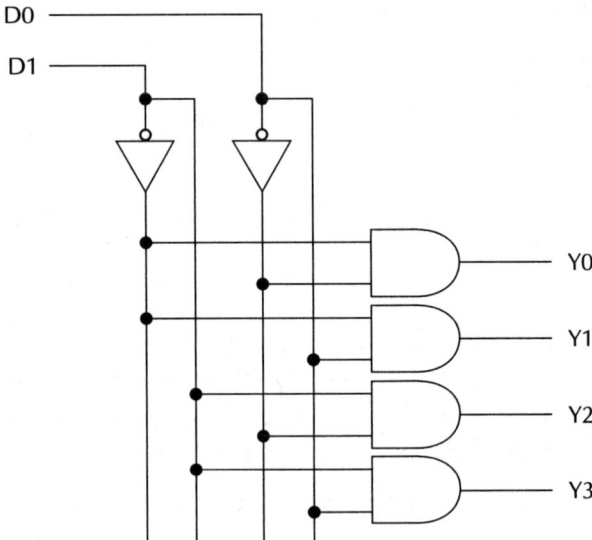

Figure 29 2-Line-to-4-Line Decoder

Solution   The circuit has two inputs and four outputs which are numerically related. We could describe the two outputs as separate names, as we could the four outputs. Or, we could show them as a related group, called a **vector**. We will show this both ways. In either case, the architecture body requires four concurrent assignments, one for each output.

Case 1 Separate Variables

```
LIBRARY ieee;
USE ieee.std_logic_1164.ALL;

ENTITY decode1 IS
  PORT (
      d1, d0       : IN  STD_LOGIC;
      y0, y1, y2, y3 : OUT STD_LOGIC);
END decode1;

ARCHITECTURE decoder1 OF decode1 IS
BEGIN
  y0 <=  (not d1) and (not d0);
  y1 <=  (not d1) and (    d0);
  y2 <=  (    d1) and (not d0);
  y3 <=  (    d1) and (    d0);
END decoder1;
```

Case 2 Vectors

```
LIBRARY ieee;
USE ieee.std_logic_1164.ALL;

ENTITY decode2 IS
  PORT (
      d  : IN  STD_LOGIC_VECTOR (1 downto 0);
      y  : OUT STD_LOGIC_VECTOR (3 downto 0));
END decode2;

ARCHITECTURE decoder2 OF decode2 IS
BEGIN
  y(0)   <=  (not d(1)) and (not d(0));
  y(1)   <=  (not d(1)) and (    d(0));
  y(2)   <=  (    d(1)) and (not d(0));
  y(3)   <=  (    d(1)) and (    d(0));
END decoder2;
```

In the latter case, we specify the length of the vector by the construct **(3 downto 0)**, indicating that Y3 is the leftmost bit in the vector. We could also use the construct **(0 to 3)**, **(4 downto 1)**, or **(1 to 4)**, depending on our requirements. A number in parentheses specifies each individual element of the vector, such as **y(3)** or **d(1)**.

We can also write the vector representation more elegantly by using a VHDL construct called a **case statement** inside a **process**, as follows:

```
LIBRARY ieee;
USE ieee.std_logic_1164.ALL;

ENTITY decode2a IS
  PORT (
      d  : IN  STD_LOGIC_VECTOR (1 downto 0);
      y  : OUT  STD_LOGIC_VECTOR (3 downto 0));
END decode2a;
```

```
ARCHITECTURE decoder OF decode2a IS
BEGIN
 decode: PROCESS (d)
 BEGIN
    CASE d IS
        WHEN "00" =>   y  <=   "0001";
        WHEN "01" =>   y  <=   "0010";
        WHEN "10" =>   y  <=   "0100";
        WHEN "11" =>   y  <=   "1000";
        WHEN others    => y   <=   "0000";
    END CASE;
 END PROCESS decode;
END decoder;
```

The **PROCESS** statement in the above VHDL code monitors the value of the input vector **d**. If **d** changes, the statements in the process are executed. The **CASE** statement evaluates **d**. For every possible combination of the 2-bit input vector, **d**, a particular value is assigned to the 4-bit vector, **y**. (For example, for the case $d_1 d_0 = 10$ (=2), the output y_2 is HIGH: $y_3 y_2 y_1 y_0 = 0100$.)

The default case ("**WHEN others =>**") is required because of the multivalued logic type STD_LOGIC_VECTOR. Since a STD_LOGIC_VECTOR can have values other than '0' and '1', the values listed for **d** don't cover all possible cases. The default output (which will never occur if we only use '0' and '1' inputs) is chosen so that no output is active in the default case. The default case would not be required if we chose to use BIT_VECTOR, rather than STD_LOGIC_VECTOR, since the listed combinations of **d** cover all possible combinations of a BIT_VECTOR.

Selected Signal Assignment

If the output of a digital circuit can be expressed as a series of alternatives, we can use a CASE statement as above, or we can use a **selected signal assignment**, to describe the circuit. This takes the form:

```
__label: WITH __expression SELECT
          __signal <= __expression WHEN __constant_value,
              __expression WHEN __constant_value,
              __expression WHEN __constant_value;
```

The signal indicated in the second line of the statement template is assigned one of several expressions, depending on the state of the expression in the first line. The constant value indicated in each line is a possible value of the expression in the first line. For example, examine the selected signal statement below:

```
circuit: WITH mode SELECT
    y    <= q   WHEN "00",
         not q  WHEN "01",
         p      WHEN "11",
         '1'    WHEN others;
```

Signal **y** is assigned one of three values, **p**, **q**, or **not q**, depending on the status of a two-bit variable called **mode**. The last clause defines a default value of logic 1.

Note that there is different notation for the logic '1' in the last clause and the values "00", "01", and "11". This has to do with the different types of the various identifiers in the statement. Identifiers **y**, **p**, and **q** are of type BIT, which can only have values '0' or '1' (single quotes). Identifier **mode** is of type BIT_VECTOR, which can be described by a **string literal** (double quotes).

Multiplexer Example

A multiplexer circuit can be easily described by a selected signal assignment statement. Recall that a multiplexer directs one of several data inputs to an output, depending on the states of one or more select inputs. An 8-to-1 multiplexer has eight data inputs and three select inputs. (Three binary inputs can be combined eight different ways.) If we label the data inputs D_0 to D_7, the selected input is the one whose decimal subscript is the same as the binary combination of select inputs $S_2S_1S_0$. For example, when $S_2S_1S_0 = 110$ (=6), input D_6 is selected. VHDL code that creates an 8-to-1 multiplexer is shown below:

```
ENTITY mux_8ch IS
  PORT (
     sel : IN   BIT_VECTOR(2 downto 0);
     d   : IN   BIT_VECTOR(7 downto 0);
     y   : OUT  BIT);
END mux_8ch;

ARCHITECTURE a OF mux_8ch IS
BEGIN
  --  Selected Signal Assignment
MUX8:    WITH sel SELECT
     y   <=    d(0)  WHEN "000",
               d(1)  WHEN "001",
               d(2)  WHEN "010",
               d(3)  WHEN "011",
               d(4)  WHEN "100",
               d(5)  WHEN "101",
               d(6)  WHEN "110",
               d(7)  WHEN "111";
  END a;
```

VHDL Counters

When using VHDL to create a counter, we can take a couple of approaches. We can use VHDL code to describe the behavior of the counter. Alternatively, we can use a predefined counter, such as those found in the MAX+PLUS II Library of Parameterized Modules (LPM) and map its ports to the ports of a VHDL design entity.

Behavioral Description of Counters

The VHDL code on the next page shows the behavioral description of a simple 8-bit counter (**ct_simp.vhd**) with asynchronous clear.

```
ENTITY ct_simp IS
    PORT(
            clk     : IN     BIT;
            clear   : IN     BIT;
            qd      : OUT    INTEGER RANGE 0 TO 255);
    END ct_simp;

ARCHITECTURE a OF ct_simp IS
 BEGIN
 -- Wait for a change on the clk or clear inputs
 PROCESS (clk, clear)
        -- Define an internal integer variable to accumulate count
        -- (Note that variable is defined outside process body)
      VARIABLE     count   : INTEGER RANGE 0 TO 255;
 BEGIN
        -- Asynchronous clear
        IF (clear = '0') THEN
            count   :=  0;
        ELSE
            -- Wait for positive clock edge
            IF (clk'EVENT AND clk = '1') THEN
                -- Increment count
                count := count + 1;
            END IF;
        END IF;
 -- Assign count value to output port
 qd   <= count;
 END PROCESS;
END a;
```

The PROCESS statement has the following syntax:

> PROCESS (*sensitivity list*)
>   [VARIABLE *variable name* : *type* [*range*];]
> BEGIN
>   *Process statements*
> END PROCESS;

Square brackets [ ] indicate an optional part of the code.

When there is a change in a variable in the sensitivity list, the process statements are executed. For a synchronous counter, the list would often include only **clock**, since any action in a synchronous circuit depends on a clock transition. Since the clear function in this counter is asynchronous, the **clear** input must also be monitored for any changes.

To hold the accumulating output value of the counter, we define a variable called **count**, presumed to have an initial value of 0, but defined for the range of 0 to 255. (This 8-bit value rolls over to 0 when the count exceeds 255.) The variable (*any* variable) is local to the process in which it is defined. We update the value of **count** by a **conditional assignment statement,** with the form:

```
IF (condition) THEN
    statement[s];
[ELSIF (condition) THEN
    statement[s];]
[ELSE
    statement[s];]
END IF;
```

The clause (IF (clear='0') THEN) monitors the asynchronous clear function independently of the clock and sets the output to 0 if true. Otherwise, the clock is monitored for a positive edge by the condition (clk'EVENT AND clk = '1'). The clause clk'EVENT (pronounced "clock tick event") is a predefined **attribute** of the clock signal and is true if there has just been a change on clock. The combination of this and the condition clk = '1' indicates that a positive edge has just occurred. If this is true, the count is incremented.

As a final step, the accumulated count must be assigned to an output port. This is done in the concurrent signal assignment qd <= count at the end of the process.

Note again the difference in types of assignments. A variable is assigned by the := operator. (e.g. count := count + 1;) A signal is assigned by the <= operator. (e.g. qd <= count).

LPM Counters in VHDL

A device from the MAX+PLUS II Library of Parameterized Modules (LPM) is specified by **ports** and **parameters**. A **port** is an input or output of the device, with a function such as clock, asynchronous clear, or asynchronous load. A **parameter** is a property of the LPM block, such as LPM_WIDTH, which specifies how many bits its parallel input or output has. Some ports and parameters, such as **LPM_WIDTH**, must be used in all instances of an LPM device. Others, such as **aclr** and **LPM_DIRECTION**, are optional.

When using an LPM counter in VHDL, we don't need to describe the behavior of the counter, since this has been done for us in the module itself. All we need to do is map the ports and parameters to the ports of the entity declaration. We do this by using a **generic map** to specify the parameters we need and a **port map** to map the ports of the LPM device either to an external port or an internal signal. The VHDL code below shows the VHDL implementation (**lpm_simp.vhd**) of the same 8-bit counter as in the previous behavioral example.

LPM components require us to use two packages: the **std_logic_1164** package in the **ieee** library and the **lpm_components** package in the **lpm** library. Since LPM components are defined using STD_LOGIC and STD_LOGIC_VECTOR types, we should use these types for our other identifiers as well.

The entity declaration defines the inputs and outputs of our counter and need not correspond to the port names for the LPM counter. That correspondence is defined in the architecture body, where we instantiate (create an instance of) the counter module. The counter is defined in a **component instantiation statement**, which takes the following form:

```
__instance_name: __component_name
    PORT MAP ( __formal_parameter => __actual_parameter,
               __formal_parameter => __actual_parameter);
```

The component name is the name of the LPM component. Formal parameters are the LPM port names. Actual parameters are the names of identifiers declared in the entity or as signals or variables.

```
-- lpm_simp.vhd
-- Eight-bit binary counter based on a component
--   from the Library of Parameterized Modules (LPM)
-- Counter has an active-LOW asynchronous clear.

LIBRARY ieee;
USE ieee.std_logic_1164.ALL;
LIBRARY lpm;
USE lpm.lpm_components.ALL;

ENTITY lpm_simp IS
    PORT (
        clk, clear  : IN    STD_LOGIC;
        qd          : OUT   STD_LOGIC_VECTOR (7 downto 0));
END lpm_simp;

ARCHITECTURE count OF lpm_simp IS
        -- Define a signal external to LPM_COUNTER for active-LOW clear
        -- (Note that signal is defined outside architecture body)
    SIGNAL clrn : STD_LOGIC;
BEGIN
        -- Instantiate 8-bit counter (no behavioral definition required)
    count8: lpm_counter
      GENERIC MAP (LPM_WIDTH  => 8)
      PORT MAP (  clock  => clk,
                  aclr   => clrn,
                  q      => qd(7 downto 0));
    clrn     <=   not clear;
END count;
```

If we want to invert the active level of an LPM input port, we must use a signal assignment statement (e.g. `clrn <= not clear;`). We need to do this because a VHDL input port cannot be "updated" (modified); only an output can be assigned a new value as a result of a Boolean expression. Thus, we create a signal called **clrn** that maps to the **aclr** (asynchronous clear) port of the LPM counter. This is connected to the **clear** input of the counter circuit via an inverter. Figure 30 shows the graphic equivalent of this mapping.

Code for two other counters is included on the following pages to further illustrate the use of VHDL. Both counters have directional control, as well as synchronous clear and load. One counter is behaviorally designed. The other is based on an LPM component.

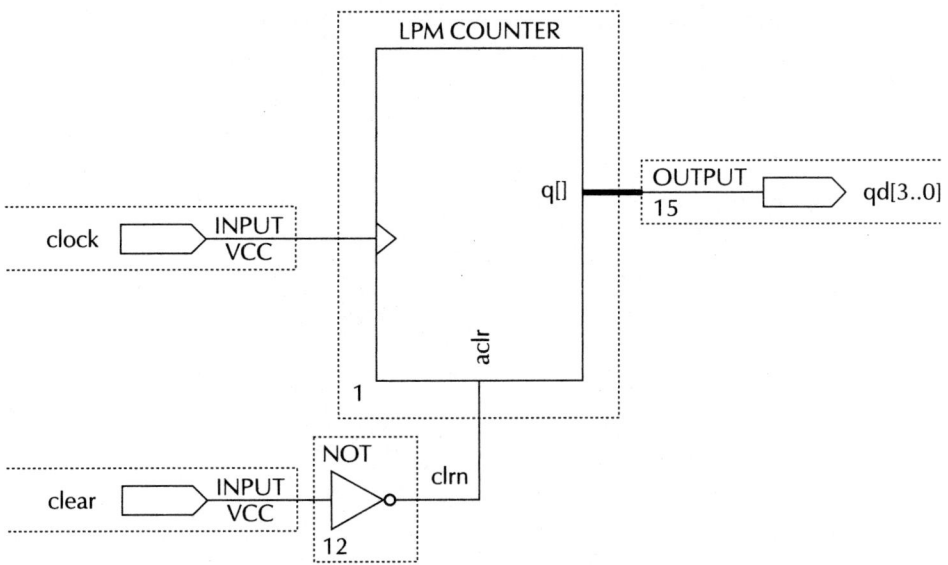

Figure 30 Equivalent Circuit of lpm_counter with Active-LOW clear

```
-- behav_ct.vhd
-- File adapted (slightly) from design example on Altera web site.
-- Behavioral description of 8-bit counter with clear, load, and
-- directional control

ENTITY behav_ct IS
PORT (
    d         : IN    INTEGER RANGE 0 TO 255;
    clk       : IN    BIT;
    clear     : IN    BIT;
    load      : IN    BIT;
    up_down   : IN    BIT;
    qd        : OUT   INTEGER RANGE 0 TO 255);
END behav_ct;

ARCHITECTURE a OF behav_ct IS
BEGIN
    -- An up/down counter
    PROCESS (clk)
        VARIABLE count : INTEGER RANGE 0 TO 255;
        VARIABLE direction : INTEGER;
    BEGIN
        IF (up_down = '1') THEN      -- Generate up/down counter
            direction := 1;
        ELSE
            direction := -1;
        END IF;

        IF (clk'EVENT AND clk = '1') THEN
            IF (load = '1') THEN          -- Generate loadable
                count := d;               -- counter.
            ELSE
                count := count + direction;
            END IF;
        END IF;
            -- The following lines will produce a synchronous
```

```
                        -- clear on the counter
                        IF (clear = '0') THEN
                            count := 0;
                        END IF;
                END IF;
                qd <= count; --Generate outputs
            END PROCESS;
        END a;


        -- count8.vhd
        -- Eight-bit binary counter based on a component
        --    from the Library of Parameterized Modules (LPM)
        -- Counter has an active-LOW synchronous reset, synchronous parallel
        -- load, and directional control

        LIBRARY ieee;
        USE ieee.std_logic_1164.ALL;
        LIBRARY lpm;
        USE lpm.lpm_components.ALL;

ENTITY count8 IS
    PORT (
        d     : IN     STD_LOGIC_VECTOR (7 downto 0);
        clk, clear, load, up_down  : IN STD_LOGIC;
        qd    : OUT    STD_LOGIC_VECTOR (7 downto 0));
END count8;

ARCHITECTURE count OF count8 IS
    SIGNAL clrn : STD_LOGIC;
BEGIN
        -- Instantiate 8-bit counter
    clock_divider: lpm_counter
                -- Parameters are specified in GENERIC MAP:
            GENERIC MAP (LPM_WIDTH => 8)
                -- Ports are specified in PORT MAP:
            PORT MAP ( clock      => clk,
                        sclr       => clrn,
                        sload      => load,
                        updown     => up_down,
                        data       => d(7 downto 0),
                        q          => qd(7 downto 0));
    clrn   <= not clear;
END count;
```

Key Terms

Architecture   A VHDL structure that defines the relationships among input, output, and internal signals or variables in a design.

Comment   Explanatory text in a VHDL (or other computer language) file that is ignored by the computer at compile time.

Concurrent   Simultaneous. At the same time.

Concurrent Signal Assignment   A relationship between an input and output signal whereby the output is updated as soon as there is a change in input, independent of the order of this and other concurrent statements. In other words, all concurrent statements are updated simultaneously.

Entity   A VHDL structure that defines the inputs and outputs of a design.

IEEE Standard 1164   The standard which defines a variety of VHDL types and operations, including the STD_LOGIC and STD_LOGIC_VECTOR types.

Instance   A single sample of a general design entity. For example, if we require several 4-bit counters in a circuit, we create several instances of a 4-bit counter.

Instantiate   Create an instance of a design entity.

Library   A collection of VHDL design units that have been previously compiled.

Package   A group of VHDL design elements that can be used by more than one VHDL file.

Port   A name assigned to an input or output of a VHDL design module.

Sequential Statements   A set of statements that are evaluated in the order in which they are written.

Syntax   The "grammar" of a computer language (i.e., the rules of construction of language statements).

Type   A set of characteristics associated with a VHDL port name, signal, or variable. Some types are BIT, BIT_VECTOR, STD_LOGIC, STD_LOGIC_VECTOR, INTEGER, BOOLEAN, and CHARACTER.

Vector   A group of digital signals or variables, usually related numerically, that can be treated as a single multibit variable (e.g., a group of signals $Y_3Y_2Y_1Y_0$ that is treated as a 4-bit variable).

VHDL (VHSIC Hardware Description Language)   An industry-standard computer language used to model digital circuits and produce programming data for PLDs.

VHSIC   Very High Speed Integrated Circuit.

Further Reading

Altera Corporation, *MAX+PLUS II VHDL Manual*

Altera Corporation, *LPM Quick Reference*

IEEE Standard VHDL Reference Manual, IEEE Press, 1993, ISBN 1-55937-376-8

Introduction to the Altera UP-1 Board

Name _____   Class _____   Date _____

Objectives   Upon completion of this laboratory exercise, you should be able to:

- Invoke MAX+PLUS II software and download an existing Programmer Object File to the Altera UP-1 Development Board.

- Download programming data to the Altera UP-1 Development Board by using a ByteBlaster cable.

- Connect switches and LEDs to inputs and outputs of a programmed logic circuit and use the switches and LEDs to determine the truth table of each programmed function.

Reference   Dueck, Robert K., *Tutorial 1—Programming CPLDs Using MAX+PLUS II*

Equipment Required   Altera   UP-1 University Lab Package:
                                        UP-1 Circuit Board
                                        ByteBlaster Download Cable
                                        MAX+PLUS II Student Edition Software
         AC Adapter, minimum output: 7 VDC, 250 mA DC
         Anti-static wrist strap
         #24 solid-core wire
         Wire strippers

Experimental Notes

Figure 31 shows a logic circuit that has been created by MAX+PLUS II. The circuit compares two binary numbers A_1A_0 and B_1B_0 and determines whether A>B, A<B, or A = B. There is an output for each of these conditions which goes LOW when the condition is satisfied. (For example: If A_1A_0 = 01 and B_1B_0 = 10, then A is less than B and the output A_LT_B goes LOW. The other outputs go HIGH.) The circuit design file can be compiled using MAX+PLUS II and the resultant programming file can be downloaded from the PC running MAX+PLUS II to a PLD chip on the Altera UP-1 Laboratory Board.

Since the purpose of this lab is to make you familiar with the Altera UP-1 board, you will not be required to enter the circuit design. You will download the compiled file to the circuit board and connect wires from the inputs and outputs of the above circuit to logic switch inputs and LED output indicators.

The four inputs can be connected to logic switches to provide the binary combinations required at the inputs of the logic circuits. The three outputs can be connected to LED indicators of the Altera UP-1 board. (Note that these indicators are active-LOW and will illuminate when its output condition (A<B, A = B, or A>B) is satisfied.)

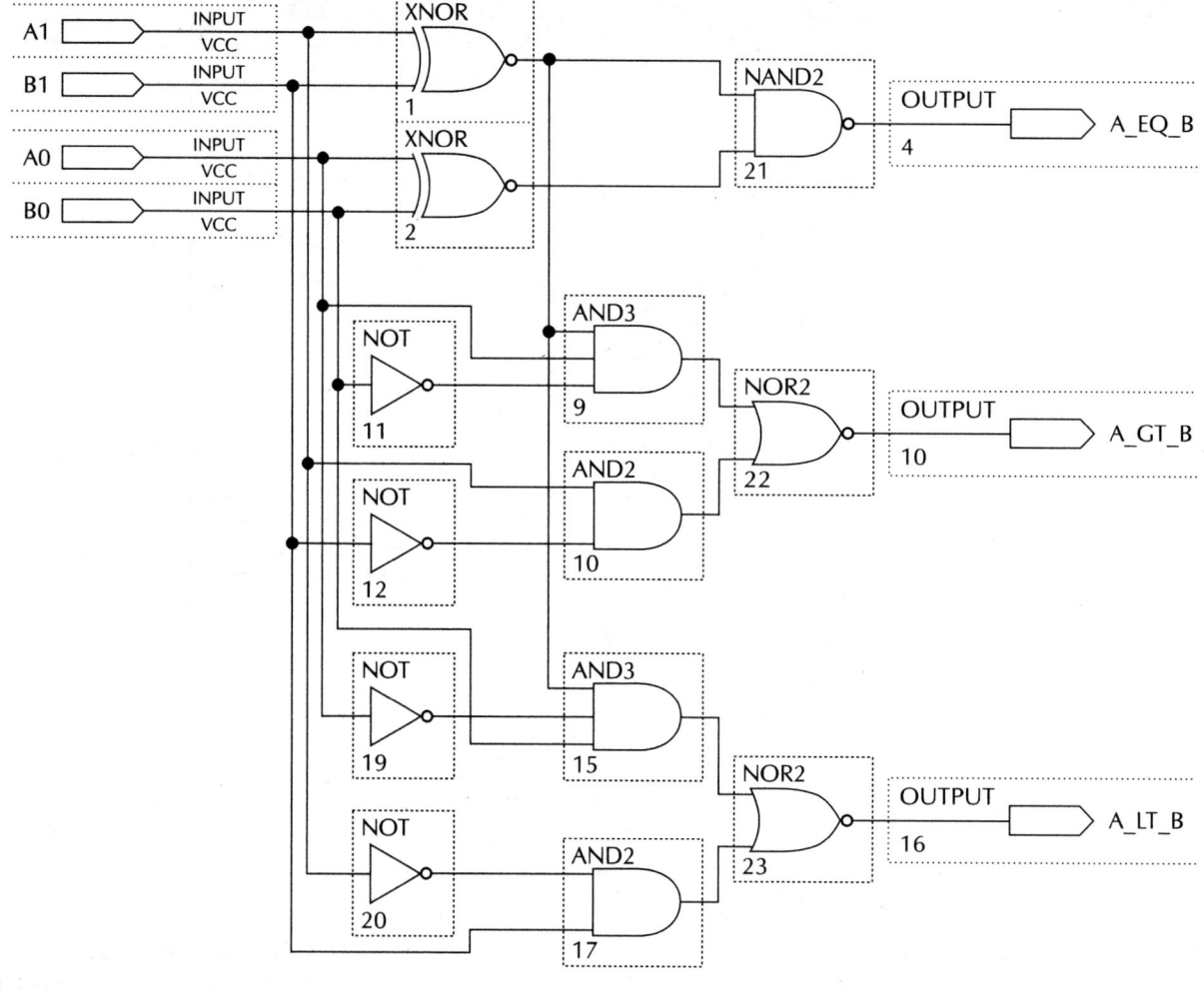

Figure 31   Graphic Design File *2bit_cmp.gdf*

Procedure

Set Up Altera Board and Download POF

1. Plug one end of the ByteBlaster cable into the parallel port (LPT1:) of the PC running MAX+PLUS II and the other end into the 10-pin male socket on the Altera UP-1 board labelled JTAG IN. You may have to use a 25-conductor cable (male-D-connector-to-female-D-connector) to make it reach. Ensure that the Altera UP-1 is powered.

2. Start the MAX+PLUS II software. Start the Programmer software, either from the MAX+PLUS II menu or from the toolbar. You should see a dialog box like the one in Figure 32.

3. From the **File** menu, choose **Select Programming File**. From the resultant dialog box, choose the file:

drive:\directory\2bit_cmp.pof

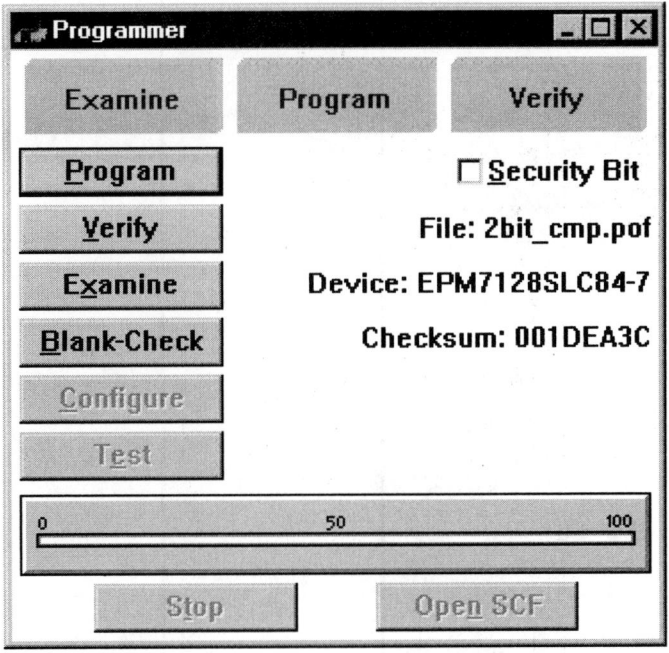

Figure 32   Programmer Dialog Box

4. Choose the **Program** button in the Programmer dialog box.

Truth Tables of Programmed Outputs

To find the truth tables of the programmed logic function outputs, you must first wire four logic switches to the inputs of your programmed device and wire an output LED to each output of the programmed CPLD. This wiring will be done between a set of pin jacks for the switches and LEDs and the four prototyping headers surrounding the EPM7128S chip on the Altera UP-1 Board. Pin assignments are shown in Table 2.

Table 2   Pin Assignments for *2bit_cmp.pof*

FUNCTION	PIN	DEVICE
A_1	12	SW1-1
A_0	16	SW1-2
B_1	15	SW2-1
B_0	17	SW2-2
A_EQ_B	4	LED1
A_GT_B	6	LED2
A_LT_B	8	LED3

When you have made the above connections, use the switches for A_1A_0 and B_1B_0 to make all binary input combinations. Fill in the truth tables for the logic functions you programmed into the EPM7128S chip. *Remember that the LEDs are active-LOW, so an illuminated LED indicates a logic LOW.*

A₁	A₀	B₁	B₀	A_EQ_B	A_GT_B	A_LT_B
0	0	0	0			
0	0	0	1			
0	0	1	0			
0	0	1	1			
0	1	0	0			
0	1	0	1			
0	1	1	0			
0	1	1	1			
1	0	0	0			
1	0	0	1			
1	0	1	0			
1	0	1	1			
1	1	0	0			
1	1	0	1			
1	1	1	0			
1	1	1	1			

Examine the completed truth table. Briefly explain to your instructor how the circuit outputs show when A = B, A>B, or A<B.

Assignment Questions

Figure 33 shows the logic diagram of a full adder. This circuit adds two input bits, A and B, as well as a carry input bit, to produce a sum bit and a carry output bit.

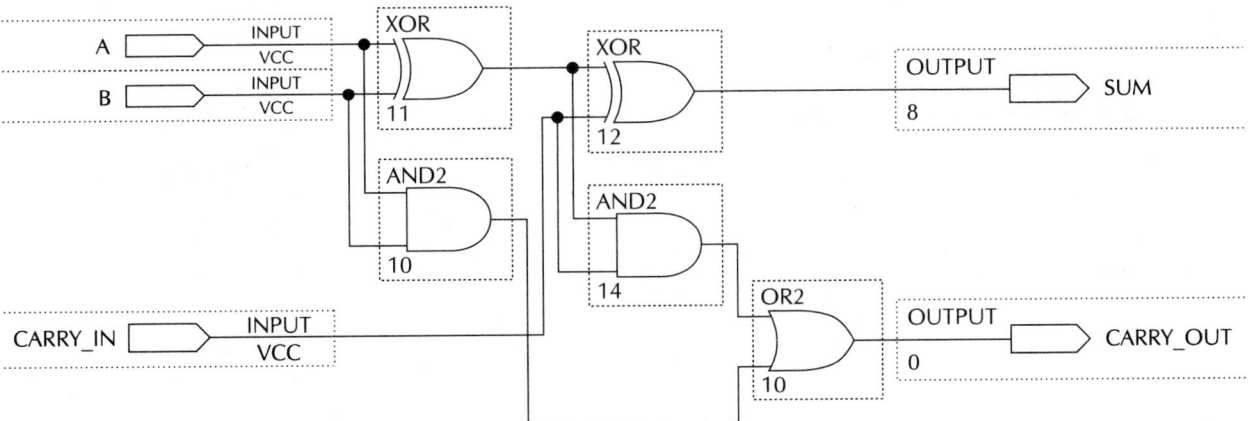

Figure 33 Full Adder

1. Write the Boolean equations for the adder circuit in Figure 33. (Use the abbreviation C_{IN} for CARRY_IN.)

2. Make a truth table for each output of the circuit.

3. From the truth tables, write the output equations in Sum-of-Products (SOP) form.

4. Simplify the SOP equations as much as possible.

5. Draw the circuit represented by the simplified SOP equations.

Seven-Segment Decoder

Name _____ Class _____ Date _____

Objectives	Upon completion of this laboratory exercise, you should be able to:

- Program an Altera EPM7128S CPLD as a hexadecimal-to-seven-segment decoder.

- Understand the usage of the VHDL selected signal assignment statement.

- Create a logic symbol from a VHDL Design File (.vhd) in MAX+PLUS II.

- Combine two seven-segment decoder symbols with a preprogrammed binary counter in a MAX+PLUS II graphic design file (.gdf) to create a two-digit hexadecimal counter.

- Assign pin numbers to the two-digit hexadecimal counter circuit, compile the design, and download it to the Altera UP-1 circuit board.

Reference   Dueck, Robert K., *Tutorial 1—Programming CPLDs Using MAX+PLUS II*
Dueck, Robert K., *Tutorial 2—Introduction to VHDL*

Equipment   Altera   UP-1 University Lab Package:
Required   UP-1 Circuit Board
                          ByteBlaster Download Cable
                          MAX+PLUS II Student Edition Software
AC Adapter, minimum output: 7 VDC, 250 mA DC
Anti-static wrist strap

Experimental Notes

Most standard seven-segment decoders are configured to generate decimal digit outputs only; there is no provision to display the hexadecimal digits A to F.

An EPM7128S CPLD can be programmed to show all hexadecimal digits on a seven-segment display as shown in Figure 34. The programming procedure involves creating a truth table for every segment of the display as a function of a 4-bit binary input.

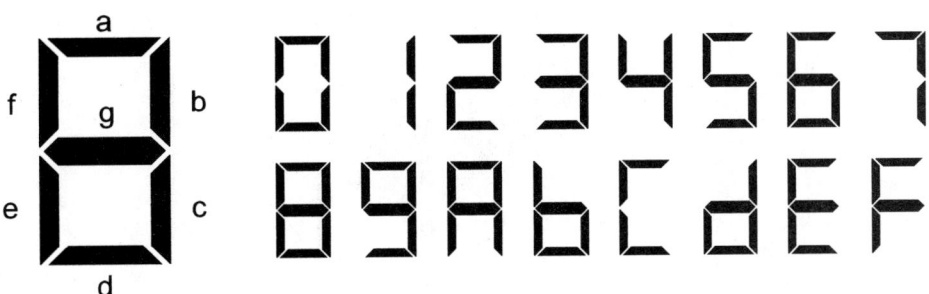

Figure 34   Hexadecimal Digit Pattern for a Seven-Segment Display

The inputs can have any value from 0000 to 1111. For each input combination, outputs for segments **a** through **g** illuminate the display segments in the required pattern.

A seven-segment decoder can be designed to drive a **common anode** or a **common cathode** display. Figure 35 shows the difference. In the common cathode display, the cathodes of all LED segments are connected together internally. This internal common point must be externally grounded for proper functioning. An LED illuminates when its anode is at a logic 1 level; the inputs are active HIGH.

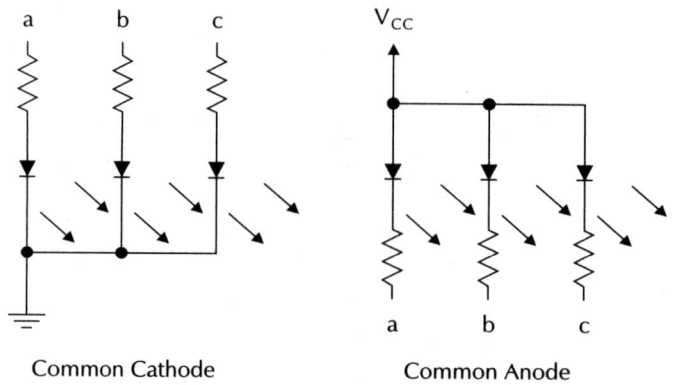

Figure 35   LED Segment Formats

The common anode display, which is the more frequently used of the two, is an active LOW device. The anodes of all LED segments are internally connected and tied to an external power supply line. Each segment is illuminated by a logic 0 at its cathode.

We can program a CPLD as a seven-segment decoder using a Hardware Description Language, such as VHDL (VHDL = VHSIC Hardware Descriptor Language; VHSIC = Very High Speed Integrated Circuit). A VHDL construct called a **selected signal assignment** can define the relationship between CPLD inputs and outputs, much as we would in a truth table.

An example of a VHDL selected signal assignment statement used like a truth table is shown for two Boolean functions, Y and Z, in Table 3.

Table 3   Functions Y and Z

D_2	D_1	D_0	Y	Z
0	0	0	0	0
0	0	1	1	1
0	1	0	1	1
0	1	1	0	1
1	0	0	0	1
1	0	1	1	1
1	1	0	1	1
1	1	1	0	1

```
-- y_and_z.vhd
-- Truth table example

ENTITY y_and_z IS
    PORT (
        d       : IN   BIT_VECTOR(2 downto 0); -- Input and output
        y, z    : OUT  BIT);
END y_and_z;

ARCHITECTURE truth_table OF y_and_z IS
    SIGNAL output: BIT_VECTOR (1 downto 0); -- Truth table output
BEGIN
    WITH d SELECT
        output  <=  "00" WHEN "000",
                    "11" WHEN "001",
                    "11" WHEN "010",
                    "01" WHEN "011",
                    "01" WHEN "100",
                    "11" WHEN "101",
                    "11" WHEN "110",
                    "01" WHEN "111";
        -- Separate the output vector to make individual pin outputs.
        y   <=   output(1);
        z   <=   output(0);
END truth_table;
```

In the file **y_and_z.vhd**, there are three inputs, D_2, D_1, and D_0, that are combined as a 3-bit input **vector** (i.e., a group of related signals). The selected signal assignment statement evaluates **d** and, based on its value, assigns a 2-bit number to a vector called **output**. The individual bits of this bit vector are split off and assigned to outputs y and z.

The section of VHDL code below contains a partial selected signal assignment statement for a common anode hexadecimal-to-seven-segment decoder. You will complete the code as part of this laboratory exercise.

There is one difference to note between the seven-segment decoder and the earlier truth table file. In the seven-segment decoder, the inputs are defined separately, not as a vector. This is so that the graphic symbol that corresponds to the file is shown with individual inputs. In order to use them in a selected signal assignment statement, we must **concatenate** (link in sequence) the inputs into a vector for internal use in the statement. Bits can be concatenated using the **&** operator.

```
-- sev_seg.vhd
-- Hexadecimal-to-seven-segment decoder

ENTITY sev_seg IS
    PORT (
        d3, d2, d1, d0    : IN BIT;   -- Use separate I/Os, not bus
        a, b, c, d, e, f, g    : OUT BIT);
END sev_seg;
```

```
ARCHITECTURE seven_segment OF sev_seg IS
-- Bit vectors for internal use
--    in decoder truth table
   SIGNAL input: BIT_VECTOR (3 DOWNTO 0);
   SIGNAL output: BIT_VECTOR (6 DOWNTO 0);
BEGIN
-- Concatenate inputs to make bit vector
   input<=   d3 & d2 & d1 & d0;
   WITH input SELECT
        output <=   "0000001" WHEN "0000",
                    "1001111" WHEN "0001",
--          .
--          .
--          .
                    "0111000" WHEN "1111",
                    "1111111" WHEN others;
     -- Separate the output vector to make individual pin outputs.
   a    <=   output(6);
   b    <=   output(5);
   c    <=   output(4);
   d    <=   output(3);
   e    <=   output(2);
   f    <=   output(1);
   g    <=   output(0);
END seven_segment;
```

Procedure

In this part of the lab, you will design a seven-segment decoder and incorporate it into the design of a two-digit hexadecimal counter. The counter is predesigned and is driven by an oscillator installed on the Altera UP-1 Development Board.

1. Write a MAX+PLUS II Text Design File (**File** menu; **New**; **Text Editor File**), using the VHDL selected signal assignment statement, to program an EPM7128S CPLD as a common anode hexadecimal-to-seven-segment decoder. Use the segment patterns shown in Figure 34. Save the file on your working drive as:

   drive:**max-plus****lab4****sev_seg.vhd**

Note   Make sure you have the correct file extension (**.vhd**). Otherwise, the file will not compile correctly.

2. From the **File** menu on MAX+PLUS II, choose **Project**, then **Set Project to Current File**.

3. From the **File** menu, choose **Create Default Symbol**. This allows you to use the seven-segment decoder as a symbol in Graphic Design File (**gdf**).

4. You will also require a file for a component called **clockdiv.** from the accompanying CD. Copy the file **clockdiv.vhd** to your working drive and place it in directory *drive:***max-plus****lab4**.

5. Create a new file and save it as **display.gdf**. From the **File** menu, choose **Project**, then **Set Project to Current File**.

6. In order to use the seven-segment decoder symbol that you created, you must also create a path to its directory by creating a User Library. From the MAX+PLUS II **Options** menu, choose **User Libraries**.

Select the *drive:*\max-plus\lab4 directory, as shown in Figure 36. Choose **Add**, then **OK**.

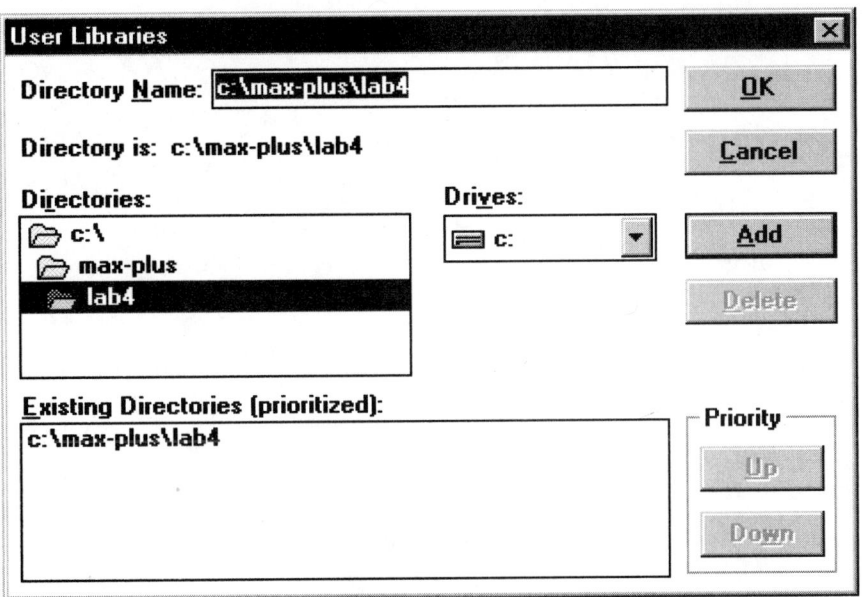

Figure 36 User Libraries Dialog

7. To enter the symbols required for the logic circuit, either use the **Symbol** menu or the right mouse button. To place the seven-segment decoder symbol, click the left mouse button to place the cursor (a flashing square) at the desired component location. Right-click the mouse to get the pop-up menu, and choose **Enter Symbol**. Select the symbol called **sev_seg** from the *drive:*\max-plus\lab4 directory.

8. Perform the following steps to create a drawing like the one in Figure 37.

a. Enter the symbol for a second seven-segment decoder.

b. Open the file **clockdiv.vhd**. Set the project to the current file and create a default symbol. Go back to **display.gdf**, set the project to the current file, and enter the symbol for **clockdiv**.

c. Use the Enter Symbol procedure to enter 16 symbols called **Output**, one for each seven-segment decoder output and one for each decimal point LED on the two-digit seven-segment display. (Type **Output** in the **Symbol Name** box of the **Enter Symbol** dialog box.)

d. The entered symbols can be positioned by first highlighting, then dragging them. Highlight a symbol by left-clicking it. Drag the symbol by holding down the mouse's left button and dragging. Position each output to line up with an output pin of one of the seven-segment decoders.

e. Connect each pin to a seven-segment output by positioning the cursor over one line
and dragging a connecting line to the other. Each time you drag a line, it can have
one right-angle bend. If a connection requires more than one bend, you will need to
draw two separate lines. Connect the 8-bit binary counter (**clockdiv**) to the inputs
of the two decoders. Finally, connect two output pins for the decimal point LEDs to
VCC. (Enter a symbol called **VCC**.)

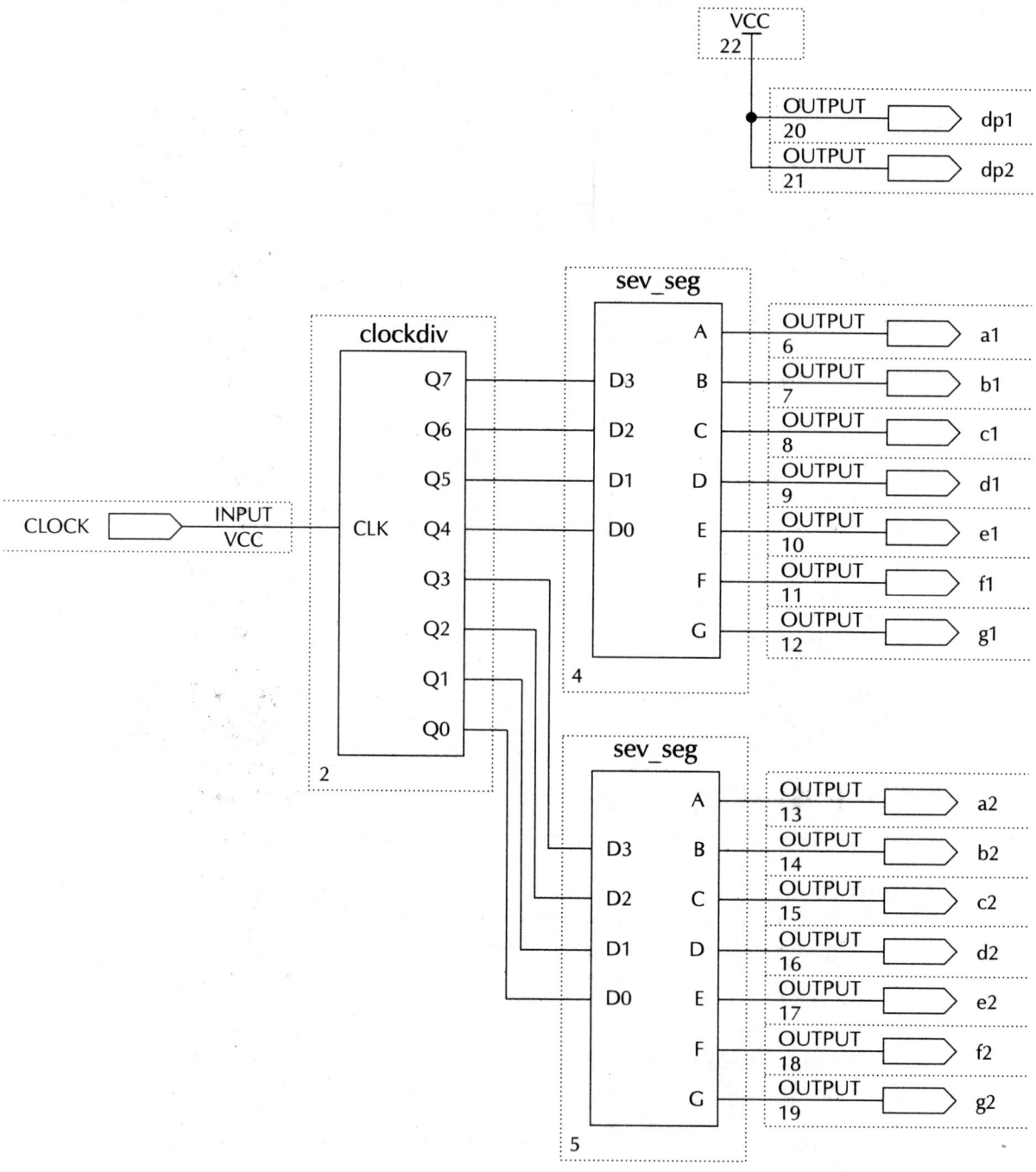

Figure 37   Two-digit Hexadecimal Counter

9. To create the hardware version of your design, you must assign the file to a specific type of device, name the pins, and assign pin numbers to the device. Since the device is hardwired into a circuit for the LED displays, we are restricted to certain pin numbers for this application.

   a. From the **Assign** menu, choose **Device**. In the Device dialog box, select the EPM7128SLC84-7 device and choose **OK**. (If you are using a MAX+PLUS II version other v7.21, you may have to select the MAX7000S family and uncheck the box that says **Show Only Fastest Speed Grades**.)

   b. Change the pin names on the outputs so that the digit 1 decoder has outputs named **a1-g1** and the digit 2 decoder has outputs named **a2-g2**, as shown in Figure 37. To change a pin name, click on the pin to highlight the whole Output primitive (component), then right-click and choose **Edit Pin Name** from the pop-up menu. Or just double-click on the pin name. The name will be highlighted, as shown in Figure 38. To replace the name, simply type the new name. Once you have entered all pin names, **Save and Compile** the design (**File, Project** menu).

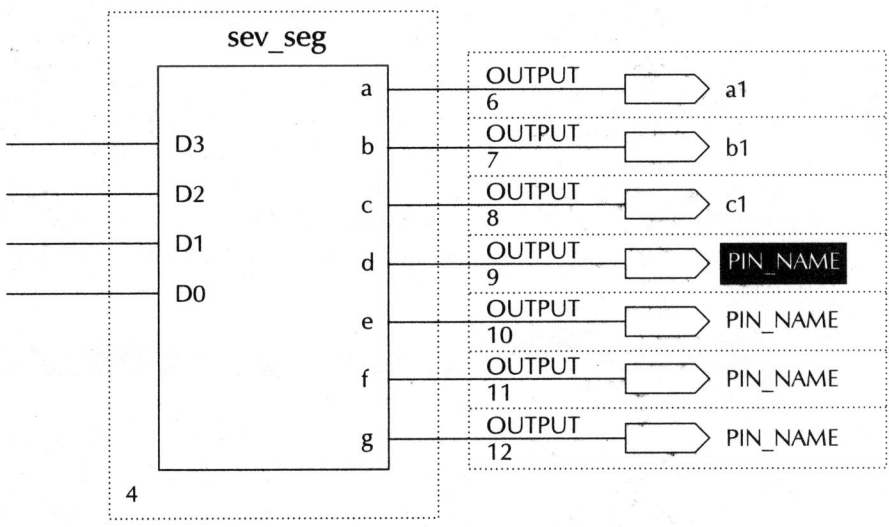

Figure 38   Assigning Pin Names

10. MAX+PLUS II is capable of creating **hierarchical designs**, which you have just done. A hierarchical design is one in which various components of a system (such as the seven-segment decoder and the binary counter) can be created separately and then brought together in a higher-level design (such as the hexadecimal counter circuit). Different levels of the hierarchy can be examined in two ways:

   a. Double-clicking on a selected component. Try this by double-clicking on one of the seven-segment decoder symbols. Describe what happens.

b. Opening the Hierarchy Display window. Do this either by selecting **Hierarchy Display** from the **MAX+PLUS II** window or clicking on the yellow pyramid icon in the MAX+PLUS II toolbar. Sketch the diagram that appears in the window.

11. Once the pin names have been assigned, you must assign the pin numbers. From the **Assign** menu choose **Pin/Location/Chip**. A dialog box appears, as shown in Figure 39. To assign a pin number, type its name in the box labeled **Node Name** and type the **Pin Number** in the appropriate box. (Follow Table 4 for assignments.) Choose **Add**. Repeat this procedure for the remaining pins. When all pins have been assigned, choose **OK**.

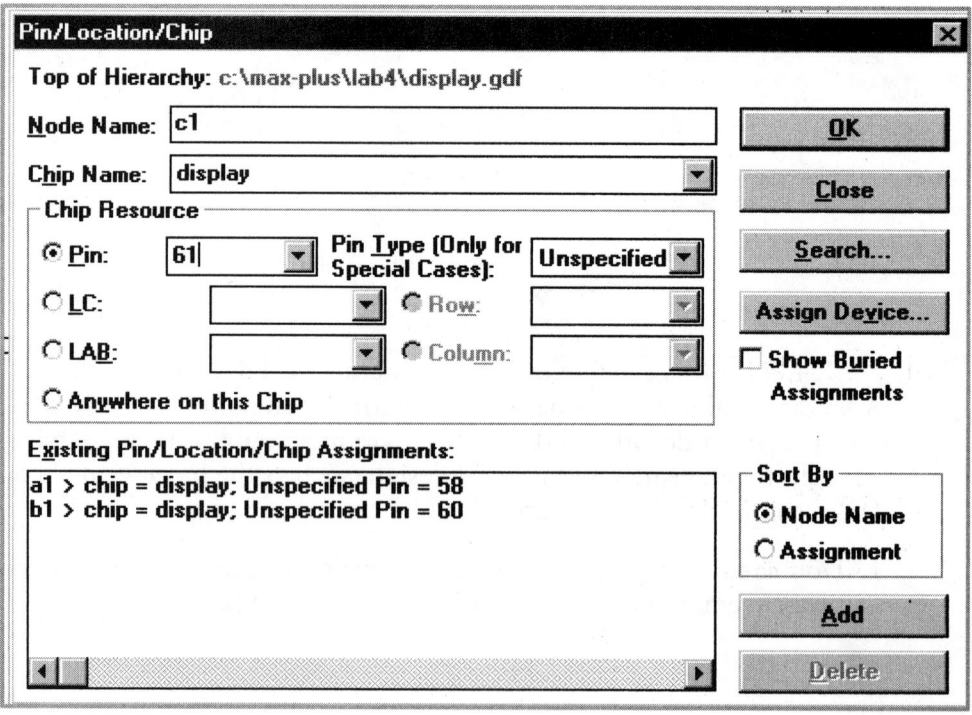

Figure 39   Assigning Pin Numbers

Table 4
EPM7128S Pin Assignments for Figure 39

Function	Pin	Function	Pin
CLK	83		
a1	58	a2	69
b1	60	b2	70
c1	61	c2	73
d1	63	d2	74
e1	64	e2	76
f1	65	f2	75
g1	67	g2	77
dp1	68	dp2	79

12. From the **File** menu, choose, **Project**, then **Save and Compile.**

13. Ensure the Altera UP-1 Board is powered and the ByteBlaster cable is plugged into the 10-pin male socket labelled JTAG IN. Select the MAX+PLUS II Programmer from the MAX+PLUS II menu or from the toolbar. Choose **Program** to download the project to the Altera UP-1 Board. When the CPLD is completely programmed, you should observe a two-digit hexadecimal count on the seven-segment display.

Changes to Symbols

If you edit **sev_seg.vhd** (or another file for which there is a circuit symbol generated), make sure to update the symbol, as follows.

Open the design file for the symbol (e.g., **sev-seg.vhd**). Set the project to the current file. From the **File** menu, choose **Create Default Symbol**. In the **gdf** file that uses it, set project to current file, highlight the symbol; then, from the **Symbol** menu, choose **Update Symbol.**

Recompile the design and download the new Programmer Object File (**pof**) to the Altera UP-1 board.

Assignment Questions

1. Modify the file **sev_seg.vhd** so that its outputs can drive a common cathode display.

2. Write a VHDL file that displays digits 0 to 9 only on a common anode display. If any other input is applied, the output should illuminate segment g only. Compile and test your file by substituting the decoders for the ones in **display.gdf**. (You can use a default case **WHEN others =>** to define anything other than the list of specified values.)

Multiplexer Applications

Name _____   Class _____   Date _____

Objectives   Upon completion of this laboratory exercise, you should be able to:

- Enter the logic circuit of a 4-to-1 multiplexer (MUX) as a Graphic Design File, using Altera's MAX+PLUS II CPLD design software.

- Create a MAX+PLUS II simulation file for the 4-to-1 multiplexer described above.

- Download the 4-to-1 MUX to a CPLD on an Altera UP-1 circuit board and test its function.

- Write VHDL design code for an 8-to-1 multiplexer, create a simulation for the design, download, and test it.

- Write VHDL code for a 4-bit-wide 2-to-1 multiplexer. Simulate and download the file. Use seven-segment input data to demonstrate the operation of the MUX.

Reference   Dueck, Robert K., *Tutorial 1—Programming CPLDs Using MAX+PLUS II*
Dueck, Robert K., *Tutorial 2—Introduction to VHDL*

Equipment Required   MAX+PLUS II Programmable Logic Development System Software
Altera UP-1 Development Board
AC Adapter, minimum output: 7 VDC, 250 mA DC
Anti-static wrist strap
ByteBlaster Download Cable

Experimental Notes

A multiplexer (abbreviated MUX) is a device for switching one of several digital signals to an output, under the control of another set of binary inputs. The inputs to be switched are called the **data inputs**; those that determine which signal is directed to the output are called the **select inputs**.

Figure 40 shows the logic circuit for a 4-to-1 multiplexer, with data inputs labelled D_3 to D_0 and the select inputs labelled S_1 and S_0. By examining the circuit, we can see that the 4-to-1 MUX is described by the following Boolean equation:

$$Y = D_0\overline{S_1}\,\overline{S_0} + D_1\overline{S_1}S_0 + D_2S_1\overline{S_0} + D_3S_1S_0$$

For any given combination of S_1S_0, only one of the above four product terms will be enabled. For example, when $S_1S_0 = 01$, the equation evaluates to:

$$Y = (D_0 \cdot 0) + (D_1 \cdot 0) + (D_2 \cdot 1) + (D_3 \cdot 0) = D_2$$

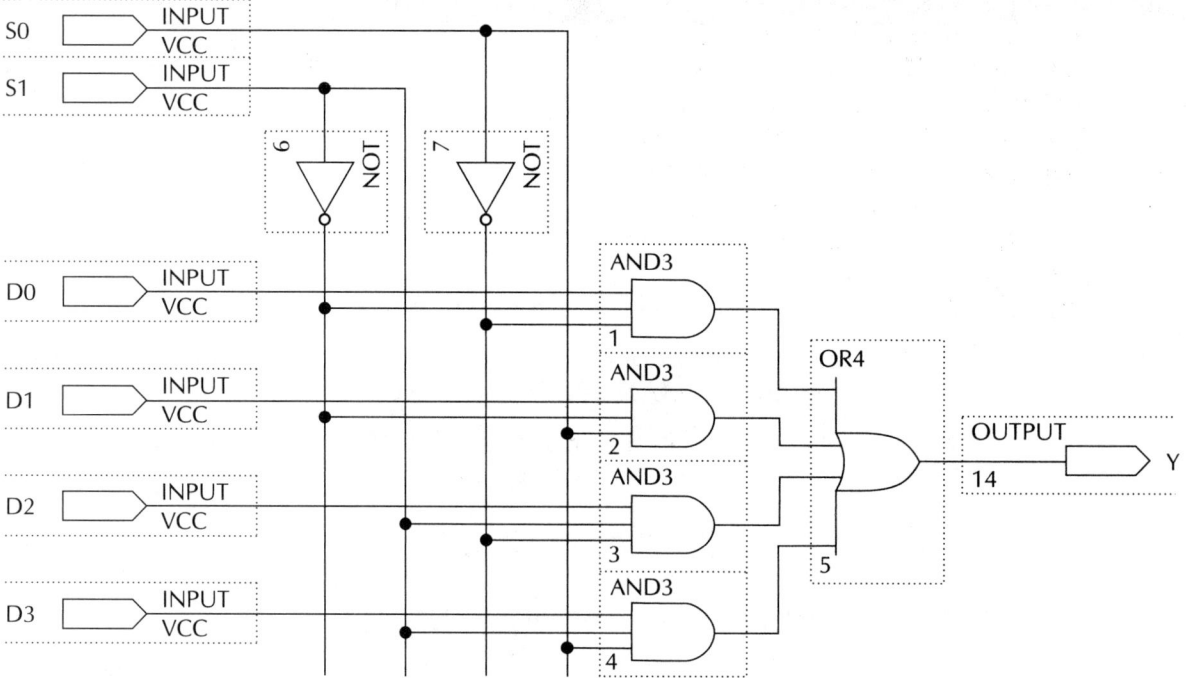

Figure 40   4-to-1 Multiplexer

The MUX equation can be described by a truth table as in Table 5. The subscript of the selected data input is the decimal equivalent of the binary combination S_1S_0.

**Table 5   4-to-1 MUX
Truth Table**

S_1	S_0	Y
0	0	D_0
0	1	D_1
1	0	D_2
1	1	D_3

Figure 41 shows two symbols used for a 4-to-1 multiplexer. The first symbol shows the individual data and select input lines. The second symbol shows the data inputs as a single 4-bit bus line and the select inputs as a 2-bit bus.

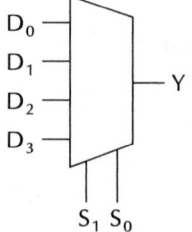

4-to-1 MUX Symbol
Showing Individual Lines

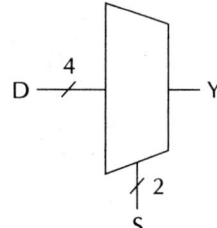

4-to-1 MUX Symbol
Showing Bus Lines

Figure 41   Multiplexer Symbols

In general, a multiplexer with n select inputs will have $m = 2^n$ data inputs. Thus, other common multiplexer sizes are 8-to-1 (3 select inputs) and 16-to-1 (4 select inputs). Data inputs can also be multiple-bit busses, as in Figure 42. In this case the select inputs switch groups of data inputs, as shown in the truth table in Table 6.

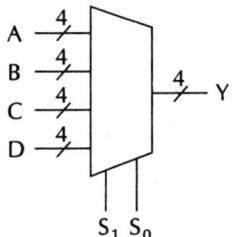

Figure 42   4-to-1 Bus MUX

Table 6   Truth Table for a 4-to-1 Bus Mux

S_1	S_0	$Y_3Y_2Y_1Y_0$
0	0	$A_3A_2A_1A_0$
0	1	$B_3B_2B_1B_0$
1	0	$C_3C_2C_1C_0$
1	1	$D_3D_2D_1D_0$

We can observe the function of a multiplexer by using time-varying waveforms, such as a series of digital pulses. If we apply a different digital signal to each data input, and step the select inputs through an increasing binary sequence, we can see the different input waveforms appear at the output in a predictable sequence, as shown by the waveforms in Figure 43.

In Figure 43, we initially see the D_0 waveform appearing at the Y output when $S_1S_0 = 00$, followed in sequence by the D_1, D_2, and D_3 waveforms when $S_1S_0 = 01$, 10, and 11, respectively. (The S_1S_0 input combination is shown as a single binary value between 0 and 3, labelled S[1..0].)

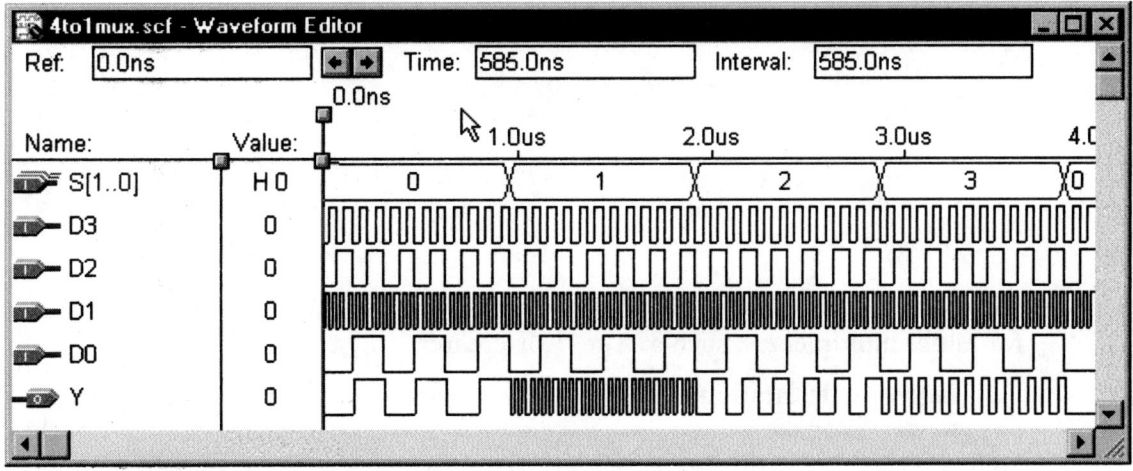

Figure 43   Switching Multiplexer Input Signals to the Output in Sequence

A multiplexer can be represented in MAX+PLUS II as a Graphic Design File, similar to the diagram of Figure 40, or in a Hardware Description Language such as VHDL.

All VHDL files require an **entity declaration**, which specifies input and output terminals, and an **architecture body** that describes the relationships between inputs and outputs. Several different VHDL constructs can be used within the architecture body to define a multiplexer. We can use a Concurrent Signal Assignment Statement, a Selected Signal Assignment Statement, or a CASE Statement within a PROCESS. We will briefly look at each form for a 4-to-1 multiplexer. Later, you will be required to extend these constructs to larger multiplexer circuits.

Concurrent Signal Assignment

This statement takes the form:

$$\_\_signal <= \_\_expression;$$

We can use this to encode the Boolean expression that describes the MUX. The VHDL file that incorporates this statement is as follows. The file name is the same as the entity name, with the file extension **vhd**. Lines of text starting with double dashes are comment lines and are for information only. They are ignored by the VHDL compiler.

```
-- mux4.vhd
-- 4-to-1 multiplexer
-- Directs one of four input signals (d0 to d3)to output,
--    depending on status of select bits (s1, s0).

-- Define inputs and outputs
ENTITY mux4 IS
  PORT(
     -- data inputs
     d0, d1, d2, d3: IN    BIT;
     -- select inputs in bus form
     s              : IN    BIT_VECTOR (1 downto 0);
     y              : OUT   BIT);
END mux4;


-- Define i/o relationship
ARCHITECTURE mux4to1 OF mux4 IS
BEGIN
  --   Concurrent Signal Assignment
  y<= ((not   s(1)) and (not   s(0)) and d0)
  or  ((not   s(1)) and (       s(0)) and d1)
  or  ((       s(1)) and (not   s(0)) and d2)
  or  ((       s(1)) and (       s(0)) and d3);
END mux4to1;
```

While the Concurrent Signal Assignment is fairly easy to use, it becomes cumbersome for larger multiplexers, such as 8-to-1 or greater.

The entity declaration will be identical for the other VHDL examples. The only change we will make will be to replace the Concurrent Signal Assignment in the architecture body with some other VHDL construct.

Selected Signal Assignment Statement

This construct has the following form:

```
__label:
WITH         __expression SELECT
 __signal <=     __expression WHEN __constant_value,
                 __expression WHEN __constant_value,
                 __expression WHEN __constant_value,
                 __expression WHEN __constant_value;
```

The 4-to-1 MUX can be described in VHDL as follows, using a Selected Signal Assignment:

```
ENTITY mux4 IS
 PORT (
     d0, d1, d2, d3 : IN   BIT;
     s              : IN   BIT_VECTOR (1 downto 0);
     y              : OUT  BIT);
END mux4;

ARCHITECTURE mux4to1 OF mux4 IS
BEGIN
M:   WITH s SELECT
     y   <= d0 WHEN "00",
            d1 WHEN "01",
            d2 WHEN "10",
            d3 WHEN "11";
END mux4to1;
```

The Selected Signal Assignment evaluates the expression in the WITH clause (in this case, the 2-bit vector, s) and, depending on its value, selects an expression to assign to y. Thus, if $s_1s_0 = 00$, $y = d_0$. If not, s_1s_0 is further evaluated. If $s_1s_0 = 01$, then $y = d_1$, and so on for the remaining values of s_1s_0.

CASE Statement within a PROCESS

In VHDL, a PROCESS is a construct containing statements that are executed if a signal in the **sensitivity list** of the PROCESS changes. The general form of a PROCESS is:

```
PROCESS (sensitivity list)
BEGIN
    statements;
END PROCESS;
```

A CASE statement can be one of the constructs used inside a process if we want to select among several alternatives. It takes the following form:

```
-- CASE statement within a PROCESS
PROCESS (__signal_name, __signal_name, __signal_name)
BEGIN
 CASE __expression IS
    WHEN__constant_value =>
          __statement;
          __statement;
    WHEN__constant_value =>
          __statement;
          __statement;
    WHEN OTHERS =>
          __statement;
          __statement;
 END CASE;
END PROCESS;
```

In our MUX example, we could use the CASE statement as follows:

```
ENTITY mux4 IS
  PORT (
     d0, d1, d2, d3  : IN     BIT;
     s               : IN     BIT_VECTOR (1 downto 0);
     y               : OUT    BIT);
END mux4;

ARCHITECTURE mux4to1 OF mux4 IS
BEGIN
-- CASE statement within a PROCESS
-- Monitor select inputs and execute if they change
     PROCESS(s)
     BEGIN
        CASE s IS
           WHEN "00"    =>  y  <=  d0;
           WHEN "01"    =>  y  <=  d1;
           WHEN "10"    =>  y  <=  d2;
           WHEN "11"    =>  y  <=  d3;
           WHEN others  =>  y  <=  '0';
        END CASE;
     END PROCESS;
END mux4to1;
```

If the select inputs change, the PROCESS statements are executed. The CASE statement evaluates the select input vector, s, and chooses a signal assignment based on its value. It is good design practice to include a default case (the "others" clause) even when there are no obvious other cases. A default case is essential when using STD_LOGIC types rather than BIT types, as '0' and '1' values do not cover all possible cases for STD_LOGIC signals. (STD_LOGIC is a nine-valued logic type, incorporating such things as "Don't Care" ('-'), "Unknown" ('X'), and "High Impedance" ('Z'), as well as '0' and '1'.)

In general, the CASE statement is a more flexible construct than the Selected Signal Assignment Statement, since the CASE can execute more than one statement for each select value. However, in the MUX definition, this doesn't matter. Both constructs achieve much the same thing.

Procedure

Graphic Design File and Simulation for 4-to-1 Multiplexer

1. Create a Graphic Design File for a 4-to-1 multiplexer as shown in Figure 40. Save the file as *drive:*\max-plus\mux_apps\4to1mux.gdf.

2. Digital circuits are often tested in the design phase, prior to committing the design to hardware, by a process called **simulation**. A simulation tool allows us to see whether the output responses to a set of circuit inputs are what we expect in our initial design idea. The simulator works by creating a timing diagram. We specify a set of input (**stimulus**) waveforms. The simulator looks at the defined relationship between input and output and creates a set of **response** waveforms.

3. Figure 43 shows a set of simulation waveforms created by the MAX+PLUS II simulation tool for the 4-to-1 MUX. Create the simulation waveforms as follows:

    From the **File** menu, select **New**. On the resultant dialog box, select **Waveform Editor File**, with a default file extension **scf**.

4. From the **File** menu, **Save As** *drive:*\max-plus\mux_apps\4to1mux.scf.

5. Specify the inputs and outputs you want to view by selecting **Enter Nodes from SNF** on the **Node** menu. In the dialog box that pops up, there are two boxes labelled **Available Nodes & Groups** and **Selected Nodes & Groups**, with an arrow (**=>**) pointing from one to the other. Select the **List** button to show the "available" signals and click the arrow to transfer them all to the "selected" box. Click **OK** to close the box.

6. Set the length of the simulation. (The default value is 1 μs, written **1.0us**.) Select **End Time** (File menu). Enter **4us**. Click **OK**.

7. Assign a set of four different waveforms to inputs D_3 to D_0. We will use four pulse waveforms (called "clock" waveforms in the simulator) of four different frequencies: 1, 2, 4, and 8 times a standard base frequency. To set the base value, select **Grid Size** in the **Options** menu and enter **20.0ns**. This is one half the desired base-value clock period, since one clock cycle takes two grid spaces.

8. Set the D_3 line to show a pulse waveform by highlighting the D_3 waveform (click in the **Value** column) and clicking the **Overwrite Clock Waveform** button on the toolbar at the left-hand side of the screen. If we want D_3 to have a frequency 2 times the base value, fill in the dialog box as shown in Figure 44 and click **OK**.

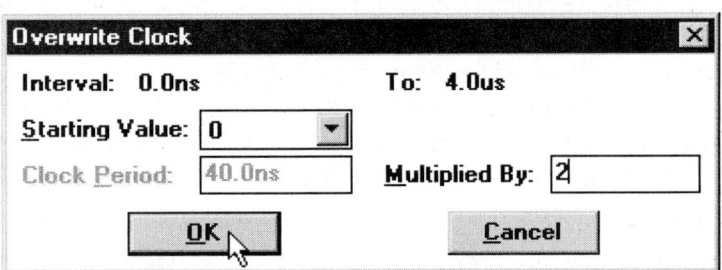

Figure 44 Overwrite Clock Dialog Box

9. Repeat the previous step for D_2 (4 times base), D_1 (1 times base), D_0 (8 times base). These frequencies were chosen to make as great a contrast as possible between adjacent inputs so that the different selected inputs could easily be seen.

10. When we created the simulation file, the select inputs were entered as separate waveforms. We can join these waveforms to make a **Group**, as shown in Figure 45, with the results indicated in Figure 43. Highlight both S_1 and S_0 by clicking on one name and dragging the mouse to the next name. From the **Node** menu, select **Enter Group** and click **OK** in the dialog box that appears.

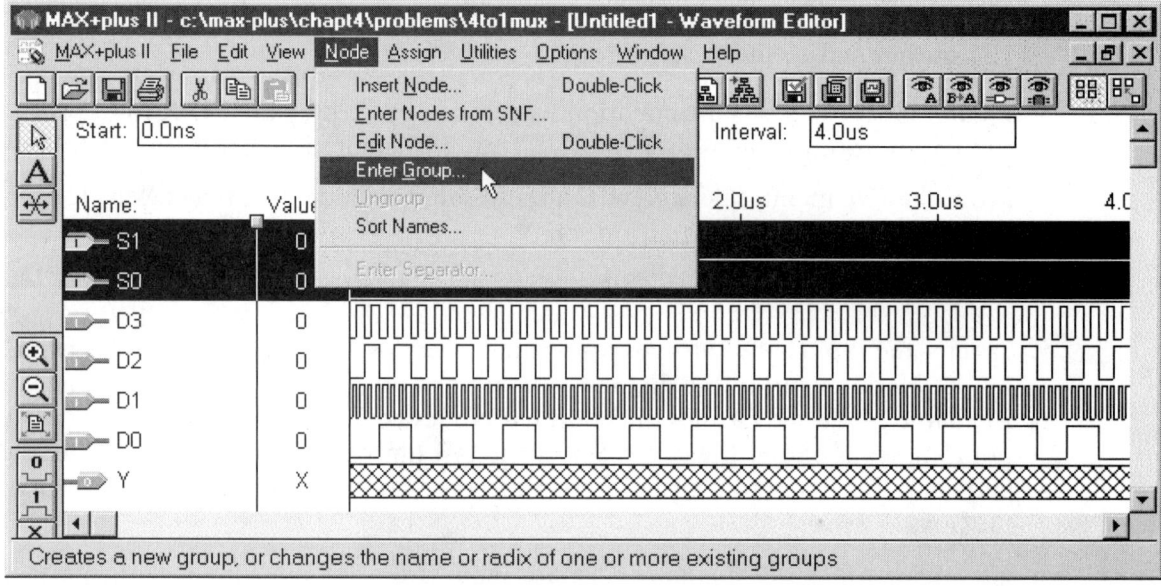

Figure 45   Making a Group Waveform

11. Use the **Overwrite Count** toolbar button to create an increasing binary count on the select input group, S[1..0]. Highlight the group by clicking in the **Value** column. Fill in the dialog box as shown in Figure 46 and click **OK**. The count is increased every 960 ns (48 × 20 ns). This allows several (at least three) cycles of every input waveform to appear at the MUX output.

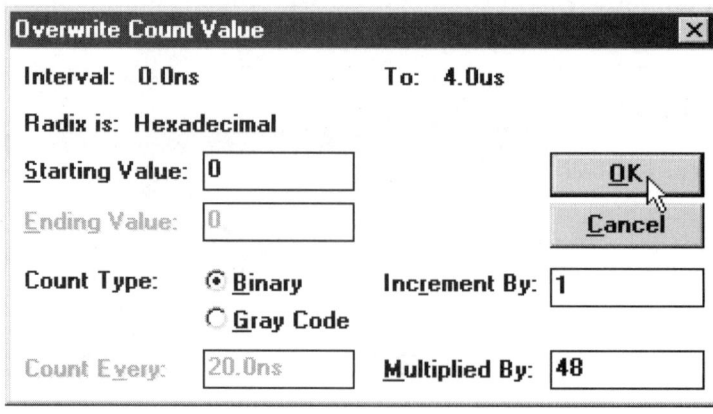

Figure 46   Entering a Binary Count on the Select Inputs

12. Save the file. From the **MAX+PLUS II** menu, bring the **Simulator** to the front and click **Start**.

13. When the simulation is finished (almost immediately), click **Open SCF** and maximize the window. From the **View** menu, select **Fit in Window**.

Testing the 4-to-1 MUX

1. One way to test the MUX function on the Altera UP-1 board is to apply a known signal to each data input, as we did in our simulation, and manually change the values of the select inputs with DIP switches. The output signal can be observed on a monitoring device such as an LED. The LED behavior will tell you which MUX input channel has been selected.

   The test circuit contains a component (**clockdiv**) that provides digital square-wave signals of binary-multiple frequencies (e.g. 3 Hz, 6 Hz, 12 Hz, 24 Hz) at different outputs. These frequencies are slow enough to be observed visually.

2. Make a default symbol (**File** menu) for the multiplexer and make sure that the folder containing the symbol is included in the list of User Libraries (**Options** menu). Create a new **gdf** file and enter the MUX symbol (**4to1mux**), along with the symbol created from the utility file **clockdiv.vhd**, as shown in Figure 47. (Recall that we used **clockdiv** in Lab 2 to drive a 2-digit hexadecimal counter. The file is stored on the accompanying CD.)

3. Save the file as *drive:***max-plus****mux_apps****mux_test.gdf**. Note that the MUX output is inverted because the LEDs on the Altera UP-1 board are active-LOW. (We could have achieved the same effect by drawing the MUX circuit with a NOR gate output.)

4. Assign a device (EPM7128SLC84-7) to **mux_test.gdf** and assign pin numbers as follows: CLK – pin 83; Y_OUT – pin 4; S_1 – pin 12; S_0 – pin 16. **Set Project to Current File** and compile the design.

5. Connect short lengths of #24 solid-core wire from the prototyping headers around the EPM7128S chip to two DIP switches (for S_1 and S_0) and to an LED (for Y_OUT). (See the Altera UP-1 User's Guide for specific pin numbering on the prototyping headers.)

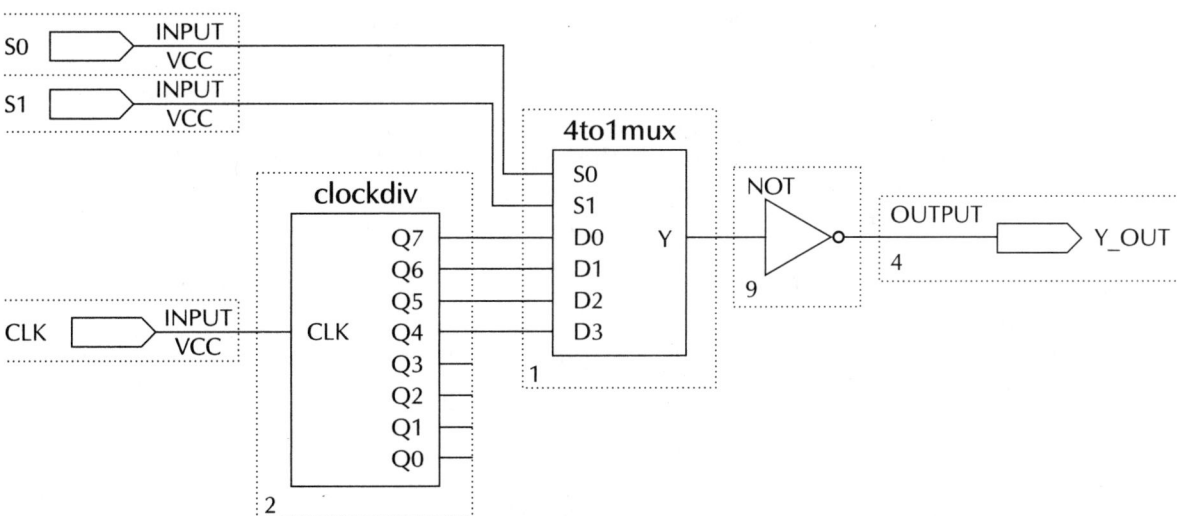

Figure 47   MUX Test Circuit

6. Download the file **muxtest.pof** to the Altera UP-1 board.

7. Set the S_1S_0 switches to 00 and observe the output of the MUX on the output LED. Repeat for values of 01, 10, and 11. Explain your observations to your instructor.

Instructor's Initials: _____

VHDL Multiplexer

1. Write a VHDL file, using a Selected Signal Assignment to design an 8-to-1 multiplexer. Define the data inputs and select inputs as BIT_VECTOR types. Save the file as *drive:*\max-plus\mux_apps\mux8_ch.vhd. Make the MUX output active-LOW so that it can drive an active-LOW LED on the Altera UP-1 board.

2. Create a simulation file for the 8-to-1 mux to verify the design operation. Show the waveforms to your instructor.

Instructor's Initials: _____

3. Make a new Graphic Design File called *drive:*\max-plus\mux_apps\mux8test.gdf, as shown in Figure 48. Note that you can connect two nodes (i.e., two points in the circuit) by labelling each node with the same name. If you want to connect individual nodes to a bus, label each node with a name and number (D0, D1, D2,..,D7) and label the bus with a group name with the range of input labels in square brackets (D[7..0]). To enter the names, highlight a node or bus and right-click the mouse. Select **Enter Node/Bus Name** from the pop-up menu that appears and enter the label text.

4. Assign a device (EPM7128SLC84-7) for the design. Refer to the table on the last page of this lab. Assign inputs D_7 to D_0 the pins indicated in the table for DIP switches SW1-1 to SW1-8. Assign the Y output to LED1 and the CLK input to pin 83. Save and compile the file.

5. Connect the D inputs on the 8-to-1 MUX to the DIP switches SW1-1 to SW1-8, using short lengths of #24 solid-core wire. Connect the Y output to LED1. Download the design to the CPLD on the Altera UP-1 board.

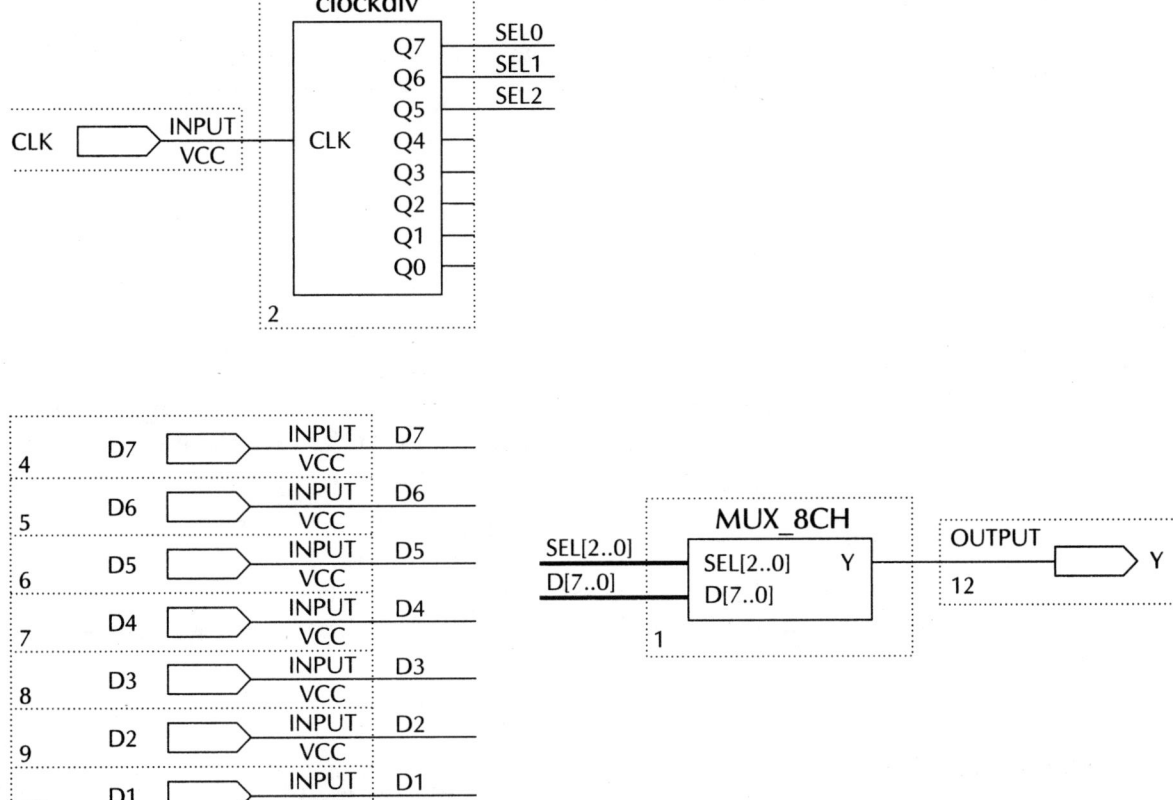

Figure 48   Test Circuit for 8-to-1 MUX

6. The circuit in Figure 48 can be used as a pattern generator; that is, a circuit to generate a repetitive flashing pattern on the output LED. The component **clockdiv** creates a repeating binary sequence from 000 to 111 on the MUX select inputs. The result is that the pattern applied to the MUX data inputs (as supplied from the DIP switch inputs) will be applied in sequence to the output LED.

7. Set the DIP switches to an alternating pattern of 0s and 1s: 01010101. What do you observe?

8. Change the pattern to 01000101. Note your observations.

9. Try other combinations such as 00110011, 00001111, 11111110 and 11111010. What do you see?

VHDL Bus Multiplexer

1. Write a VHDL file that defines a multiplexer that switches two 4-bit inputs, **x** and **y** to a 4-bit output, **z**. Define **x**, **y**, and **z** as type BIT_VECTOR. Save the file as *drive:*\max-plus\mux_apps\quad2to1.vhd.

2. Create a simulation for the Quadruple 2-to-1 MUX.

3. Create a default symbol for the MUX and, in a new **gdf**, connect the **z** outputs to a hexadecimal-to-seven-segment decoder created in Lab 2, as shown in Figure 49.

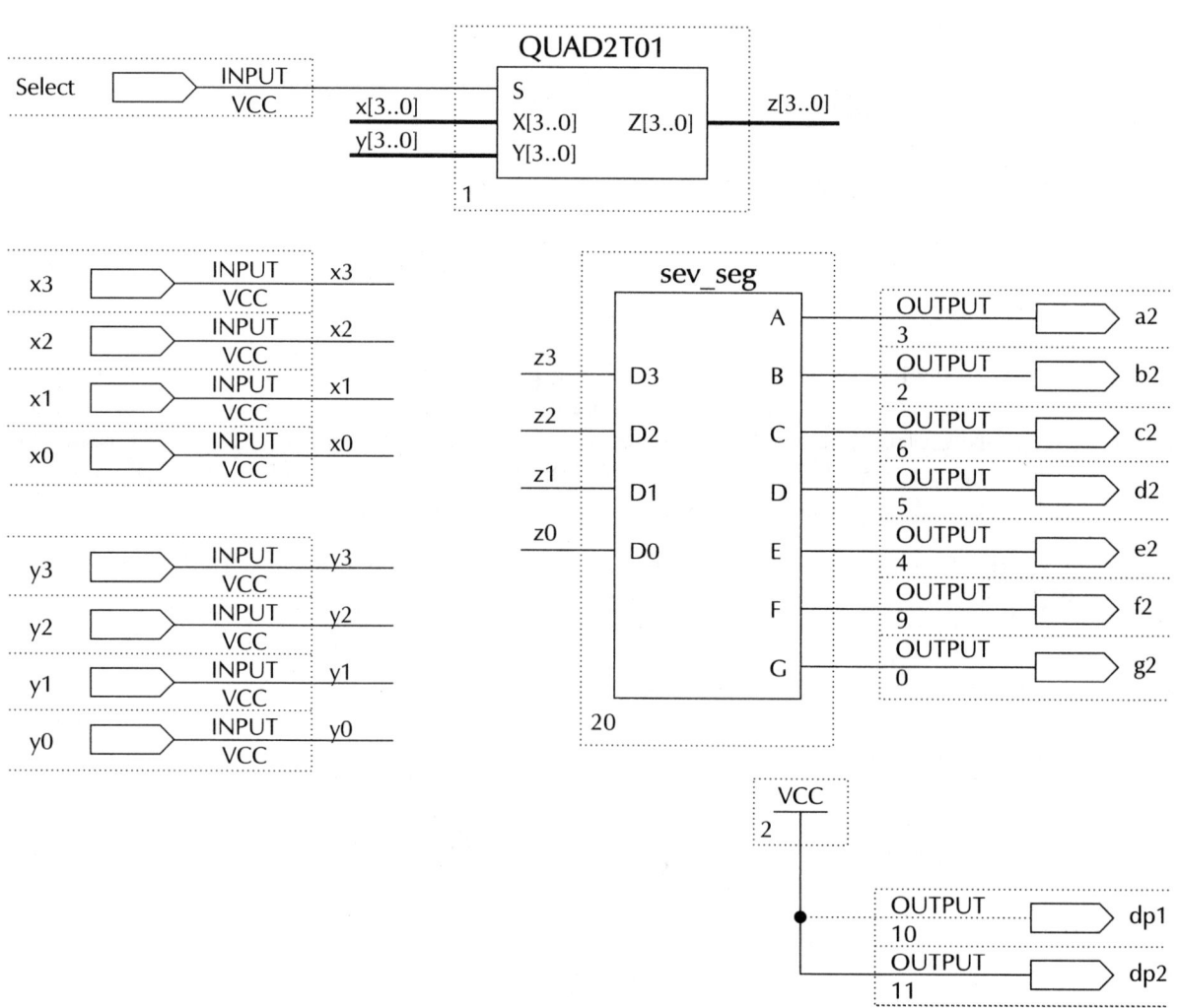

Figure 49   Test Circuit for a Quad 2-to-1 MUX

4. Assign inputs **x[3..0]** to the pins for SW1-1 to SW1-4. Assign inputs **y[3..0]** to the pins for SW1-5 to SW1-8. Assign the Select input to the pin for DIP switch SW2-8. Assign the seven-segment outputs as shown on the Pin Assignments table at the end of this lab.

5. Save and compile the file and download it to the Altera UP-1 board. Connect the pin jacks for the DIP switches to the prototyping headers for the EPM7128S chip on the Altera UP-1 board.

6. Set switches SW1-1 to SW1-4 (**x**) to 0101. Set switches SW1-5 to SW1-8 (**y**) to 1010. What effect does the **Select** switch have on the digit shown on the seven-segment display? Show the results to your instructor.

Instructor's Initials: _____

EPM7128LC84-7 Pin Assignments
Altera UP-1 Board

SEVEN SEGMENT DIGITS			
Function	**Pin**	**Function**	**Pin**
a1	58	a2	69
b1	60	b2	70
c1	61	c2	73
d1	63	d2	74
e1	64	e2	76
f1	65	f2	75
g1	67	g2	77
dp1	68	dp2	79

PUSHBUTTONS			
Function	**Pin**	**Function**	**Pin**
PB1	51	PB2	52

EPM7128LC84-7 Pin Assignments
Altera UP-1 Board (*continued*)

DIP SWITCHES			
Function	**Pin**	**Function**	**Pin**
SW1-1	12	SW2-1	15
SW1-2	16	SW2-2	17
SW1-3	18	SW2-3	21
SW1-4	20	SW2-4	25
SW1-5	22	SW2-5	27
SW1-6	24	SW2-6	29
SW1-7	28	SW2-7	31
SW1-8	30	SW2-8	33

LED OUTPUTS			
Function	**Pin**	**Function**	**Pin**
LED1	4	LED9	81
LED2	6	LED10	5
LED3	8	LED11	9
LED4	10	LED12	11
LED5	56	LED13	50
LED6	57	LED14	48
LED7	54	LED15	46
LED8	55	LED16	44

Unassigned: Pins 34, 35, 36, 37, 39, 40, 41, 45, 49, 80

Counters and Decoders

Name _____ Class _____ Date _____

Objectives Upon completion of this laboratory exercise, you should be able to:

- Enter the design for a 3-bit synchronous binary counter as a MAX+PLUS II Graphic Design File.

- Create a simulation file to verify the operation of the counter design.

- Incorporate the counter into a circuit that displays an increasing count as a decimal digit on a seven-segment display.

- Download the counter's Programmer Object File to the Altera UP-1 board and demonstrate the circuit function, using Clock and Reset inputs from the Altera UP-1 Board.

- Modify the counter circuit to provide binary-decoded outputs.

- Simulate the counter and decoder circuit.

- Program the CPLD on the Altera UP-1 board with the decoded counter design.

- Wire the Altera UP-1 Board to display decoded outputs and demonstrate the circuit operation.

Reference Dueck, Robert K., *Tutorial 1—Programming CPLDs Using MAX+PLUS II*
Dueck, Robert K., *Tutorial 2—Introduction to VHDL*
Dueck, Robert K., *Design Case Study—Switch Debouncer for the Altera UP-1 Circuit Board*

**Equipment
Required** Altera UP-1 University Lab Package:
UP-1 Circuit Board
ByteBlaster Download Cable
MAX+PLUS II Student Edition Software
AC Adapter, minimum output: 7 VDC, 250 mA DC
Anti-static wrist strap
#24 solid-core wire
Wire strippers

Procedure

Three-Bit Synchronous Counter

Figure 50 shows the logic diagram of a 3-bit synchronous counter with asynchronous reset.

The counter output, $Q_2Q_1Q_0$, progresses in a repeating binary sequence from 000 to 111, advancing by 1 with every positive edge of the CLOCK waveform. The output is immediately set to 000 when the RESET input goes LOW.

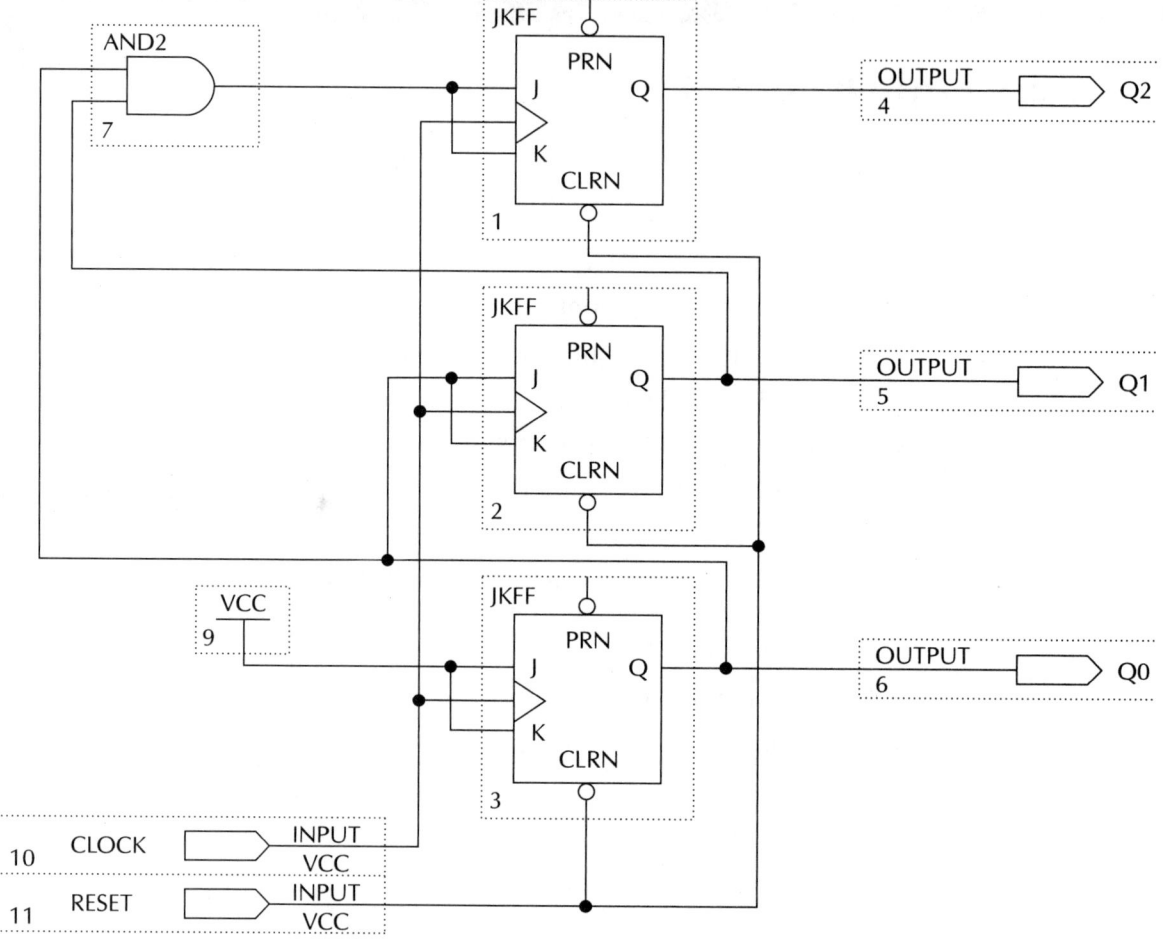

Figure 50   Three-Bit Synchronous Counter

1. Enter the logic diagram of Figure 50 as a Graphic Design File in MAX+PLUS II. **Save As 3bit_ctr.gdf** in a new folder on your working drive. From the **File** menu, select **Project** and **Set Project to Current File.** It is not necessary to assign pin numbers at this time.

2. Normally, MAX+PLUS II assumes that synchronous devices, such as flip-flops, will be clocked by a dedicated global clock, that is, a clock signal applied to all devices within the CPLD from a specially assigned pin. Since we will eventually require a switch debouncer circuit to manually clock the flip-flops, we must disable the global Clock and Clear functions. Do this as follows:

   From the **Assign** menu, select **Global Project Logic Synthesis.** In the dialog box that comes up, look for the box labeled **Automatic Global.** Uncheck the entries labeled **Clock** and **Clear.**

3. From the **File** menu, select **Create Default Symbol.**

4. Compile the design, using the MAX+PLUS II compiler.

Creating a Simulation File for the 3-Bit Counter

Figure 51 shows a set of simulation waveforms created by the MAX+PLUS II simulation tool for the synchronous 3-bit counter. The Reset input is set to logic 1 to disable it. The Clock input pulses once every 40 ns. The Q outputs respond by showing a binary count waveform. If you read the values of $Q_2Q_1Q_0$ in a vertical line, you will see that they progress in a repeating binary sequence: 000, 001, 010,...,111, 000,...

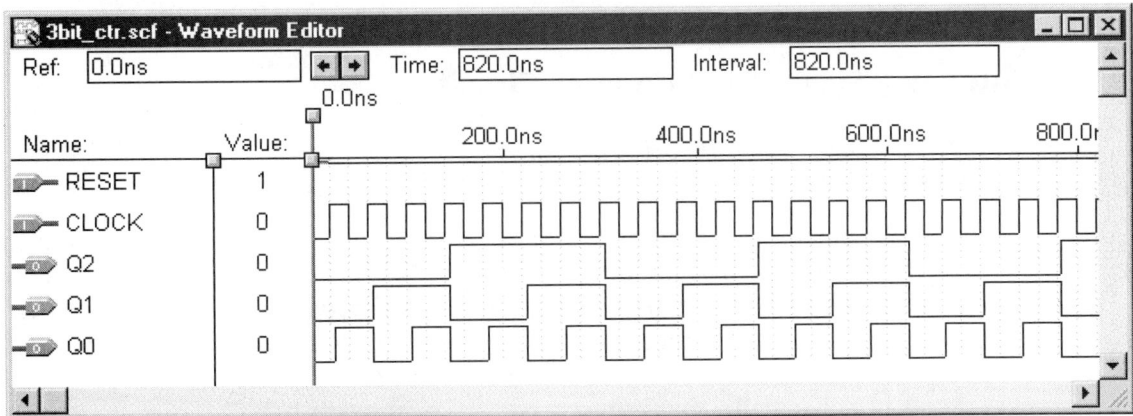

Figure 51   Simulation Waveforms for a 3-Bit Counter

Create the simulation waveforms as follows:

1. From the **File** menu, select **New**. On the resulting dialog box, select **Waveform Editor File**, with a default file extension **scf**.

2. From the **File** menu, **Save As 3bit_ctr.scf** in the same folder as **3bit_ctr.gdf**.

3. Specify the inputs and outputs you want to view by selecting **Enter Nodes from SNF** on the **Node** menu. Click **OK** when inputs and outputs have been selected.

4. Set the Reset line to logic 1. To do this, click in the **Value** column beside **Reset** to highlight the line and click the button on the left side of the screen that shows the 1 and the pulse.

5. Set the **Grid Size** (**Options** menu) to 20 ns. Make sure that **Snap to Grid** is checked.

6. Set the Clock line to show a pulse waveform by highlighting the Clock waveform (click in the **Value** column) and clicking the **Overwrite Clock Waveform** button on the toolbar at the left-hand side of the screen. Click **OK** in the resultant dialog box.

7. **Save** the file. From the **MAX+PLUS II** menu, bring the **Simulator** to the front and click **Start**.

8. When the simulation is finished, click **Open SCF** and maximize the window. From the **View** menu, select **Fit in Window**.

Show the result to your instructor and explain the waveforms.

Instructor's Initials: _____

The waveforms in your simulation do not test the operation of the Reset function. Edit the waveform file you created to add a LOW-going pulse to the Reset line. Select the portion of the Reset line from 400 ns to 420 ns by dragging the mouse at that point in the waveform. Click the toolbar button showing the 0 pulse. **Save** the file and run the simulator again.

Show the result to your instructor and explain the new waveforms.

Instructor's Initials: _____

Entering a Single-Digit Counter Circuit

1. Create a new **gdf** file and save it on your working drive as **count7.gdf**. Set the project to the current file.

2. Enter the schematic shown in Figure 52, using your counter symbol created earlier in this lab and the seven-segment decoder from Lab 2, as well as the switch debouncer circuit found in the file **debounce.vhd** on the accompanying CD. For a further description of the debouncer circuit, see *Switch Debouncer for the Altera UP-1 Circuit Board*.

3. Assign the pin numbers as shown in Table 7. **Set Project to Current File** and **Save and Compile** the project.

Table 7   **Pin Assignments for Single-Digit Counter**

FUNCTION	PIN	FUNCTION	PIN
SYSCLOCK	83	CLOCK	51
RESET	52		
		a2	69
		b2	70
		c2	73
		d2	74
		e2	76
		f2	75
g1	67	g2	77
dp1	68	dp2	79

4. Download the file **count7.pof** to the EPM7128S chip on the UP-1 board.

Testing the Single-Digit Counter

You will use the pushbutton switches on the Altera UP-1 Board to test the Single-Digit Counter.

1. Using a length of #24 wire, connect MAX_PB1, the terminal of one of the pushbuttons on the UP-1 Board, to pin 51 on the MAX Prototyping Header. This is the Clock input.

2. Connect MAX_PB2 to pin 52 for the Reset function.

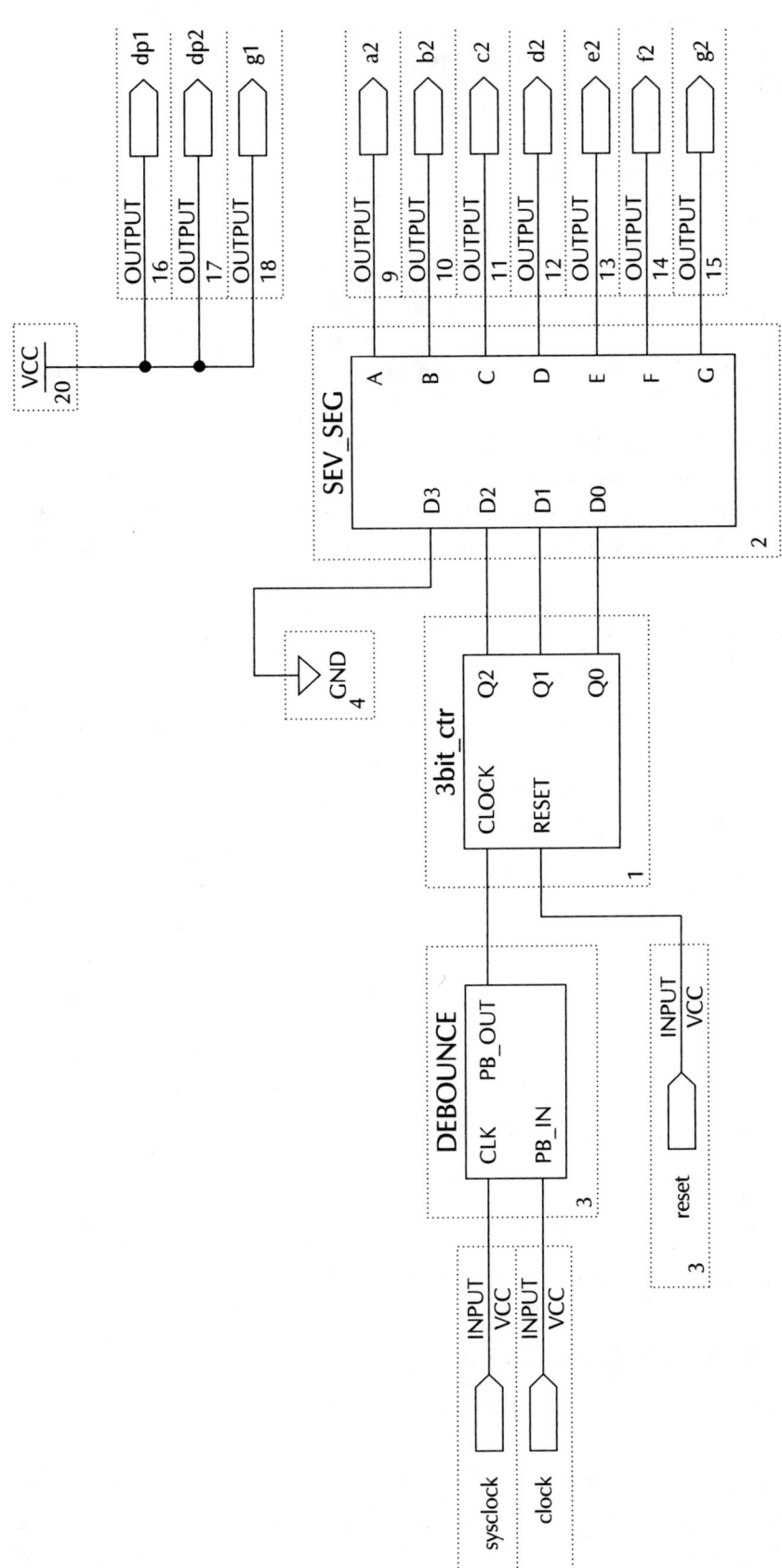

Figure 52   Single-Digit Counter

3. Press the Clock pushbutton a few times and observe the operation of the seven-segment display on the Altera UP-1 board. Press the Reset pushbutton and observe the response. Demonstrate the operation of the circuit to your instructor.

Instructor's Initials: _____

Binary Decoder for a Counter Output

A binary decoder is used to detect the various counter states as they appear at the counter's Q outputs. (For example, when the counter output is 011 (=3), only decoder output Y3 will go HIGH. At all other times Y3 is LOW.) Figure 53 shows a binary decoder added to our single-digit counter circuit. Since an n-bit counter has 2^n states, a binary counter decoder will have n inputs and up to 2^n outputs. Our 3-bit counter can be decoded by a 3-line-to-8-line decoder. Each decoder output will drive an LED. When the counter is clocked, the LEDs will light in sequence, one at a time.

Binary decoders are easily described by Boolean equations. A 3-line-to-8-line decoder will have eight defining Boolean equations, each in terms of the three decoder inputs. For example, the output for Y0 in Figure 53 (which responds to state 000) can be described by the equation:

$$\overline{Y0} = \overline{Q_2}\,\overline{Q_1}\,\overline{Q_0}$$

The output is indicated as active-LOW, since the UP-1 LEDs are active LOW.

1. Write a VHDL File for the binary decoder shown in Figure 53. The file should consist of an ENTITY declaration and an ARCHITECTURE body. (Use the VHDL Template menu as a basis for these components, if you wish.) The ARCHITECTURE body can be created with eight Boolean equations written as Concurrent Signal Assignment Statements.

   The equation for Y0 is:

   $$Y0 <= not((not\ Q_2)and(not\ Q_1)and(not\ Q_0));$$

   The VHDL file can also define the decoder with a Selected Signal Assignment Statement or a CASE statement.

2. Save the file as **decode8.vhd** on your working drive. Set the project to the current file and from the **File** menu, select **Create Default Symbol**.

3. Add the decoder symbol (decode8) to the schematic for the single-digit counter. Save the modified file as *drive:\your directory*\ct7_dcd.gdf and set the project to the current file. Assign additional pin numbers as shown in Table 8.

Table 8   Pin Assignments for Binary Decoder

FUNCTION	PIN	FUNCTION	PIN
Y0	4	Y4	54
Y1	6	Y5	55
Y2	8	Y6	56
Y3	10	Y7	57

4. Compile the design.

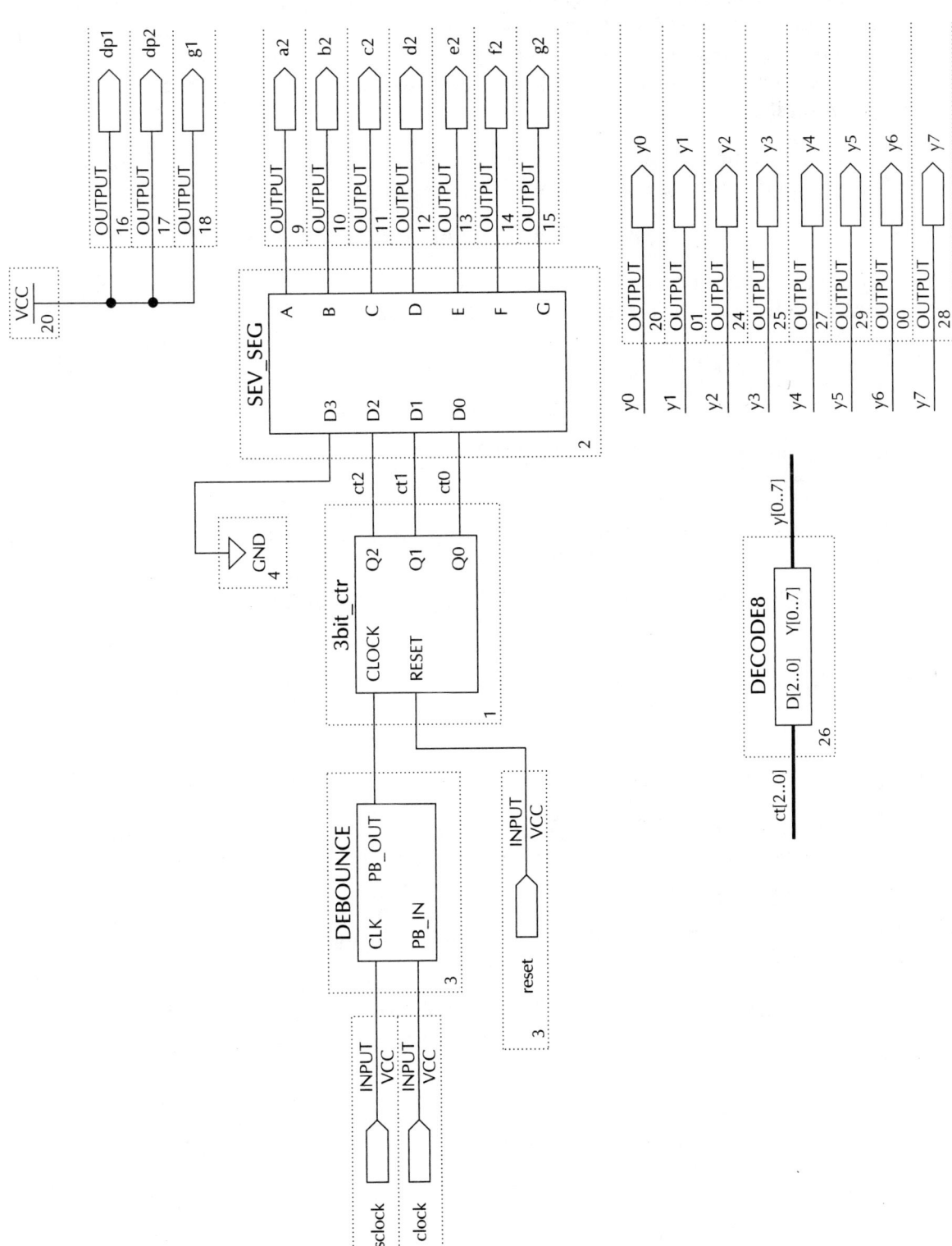

Figure 53 Counter with Seven-Segment and Binary Decoders

Creating a Simulation File for the Binary Decoder

1. In order to create a simulation for this circuit, you will need to save the file under another name (e.g., **ct7_dcd1.gdf**) and remove the switch debouncer from the circuit. (The debouncer has a 16-bit clock divider inside it; to generate *one* output pulse from the debouncer requires $2^{16} = 65,536$ input pulses. This is not practical to simulate for a simple application such as ours.) **Set Project to Current File** and select **Project Save and Compile**.

2. From the **File** menu, select **New**. On the resultant dialog box, select **Waveform Editor File**, with a default file extension **scf**.

3. From the **File** menu, **Save As ct7_dcd1.scf** in the folder on your working drive.

4. Specify the inputs and outputs you want to view by selecting **Enter Nodes from SNF** on the **Node** menu. Select **List** and click the arrow (=>).

5. Set the Reset line to logic 1.

6. Set the Clock line to show a pulse waveform.

7. Save the file. From the **MAX+PLUS II** menu, bring the **Simulator** to the front and click **Start**.

8. Click **Open SCF** and maximize the window. From the **View** menu, select **Fit in Window**.

   Show the result to your instructor and explain the waveforms.

Instructor's Initials: _____

Testing the Binary Decoder

1. Using short lengths of #24 wire, connect the pins and connectors on the Altera UP-1 Board as shown in Table 8. These connections join the binary decoder outputs to the LEDs on the Altera UP-1 Board.

2. Download the compiled file for the counter and decoders (**ct7_dcd.pof**) to the MAX7000S chip on the Altera UP-1 board.

3. Press the Clock pushbutton a few times and observe the operation of the LED outputs on the Altera UP-1 Board. Press the Reset pushbutton. Demonstrate the operation of the circuit to your instructor.

Instructor's Initials: _____

Assignment Questions

1. **a.** The CLOCK pushbutton on the Altera UP-1 Board is debounced by a circuit programmed into the MAX7000S chip. Briefly explain what that means and how an undebounced pushbutton would affect the operation of the 3-bit counter if it were used as a CLOCK input. Use a timing diagram in your explanation.

   b. Draw a circuit, based on a NAND latch, that would remove switch bounce from a circuit. Give a brief explanation of how it works.

2. a. Write the Boolean equations for the J and K inputs of the flip-flops used in the 3-bit counter shown in Figure 50.

    b. Write the J and K equations for a 4-bit synchronous counter.

    c. Draw the circuit for a 4-bit synchronous counter based on JK flip-flops. The counter should have an asynchronous reset.

Parameterized Counters and Shift Registers

Name _____ Class _____ Date _____

Objectives Upon completion of this laboratory exercise, you should be able to:

- Design, program and test a CPLD-based binary counter, using Altera's MAX+PLUS II software for schematic entry.

- Design, program and test a CPLD-based shift register, using the schematic entry function of Altera's MAX+PLUS II software.

- Use the Library of Parameterized Modules (LPM) in MAX+PLUS II to implement counter and shift register functions.

Reference Altera Reference Manual, *MAX+PLUS II Getting Started*
Altera, *LPM Quick-Reference Guide*
Dueck, Robert K., *Tutorial 2—Introduction to VHDL*

Equipment Altera UP-1 University Lab Package:
Required   UP-1 Circuit Board
        ByteBlaster Download Cable
        MAX+PLUS II Student Edition Software
    AC Adapter, minimum output: 7 VDC, 250 mA DC
    Anti-static wrist strap
    #24 solid-core wire
    Wire strippers

Experimental Notes

Altera's MAX+PLUS II PLD design and programming software allows us to enter digital designs in a variety of different ways.

We can use **schematic entry**, where we draw the circuit with a CAD (Computer Aided Design) tool and allow the PLD (Programmable Logic Device) design and programming software to convert it into a form that can be downloaded into a PLD. As we have seen in previous labs, this is useful for fairly small circuits or for connecting complex components as part of a **hierarchical design**, but is inefficient to use for circuits requiring large numbers of circuit interconnections or many inputs and outputs. (A hierarchical design is one that is ordered in layers or levels. The highest level of design contains components that are themselves complete designs. These components may, in turn, have lower level designs embedded within them.)

We can also enter design information using a **hardware description language (HDL)**, such as Altera Hardware Description Language (**AHDL**) or VHSIC Hardware Description Language (**VHDL**). (VHSIC = Very High Speed Integrated Circuit.) Both these languages allow us to specify the operation of a circuit by writing a text-based description that conforms to certain syntax rules of the language. Even though it does take some work to learn the language rules, it is ultimately easier to use this method to specify complex circuitry. If necessary, an HDL file can be represented as a graphic circuit symbol that is used in a higher level of a hierarchical design.

MAX+PLUS II offers a library of components that can be used in either an HDL or schematic design. These **LPM (Library of Parameterized Modules)** components can be modified easily to create designs of any required size. For example, a parameter called LPM_WIDTH can be set to any value to make a counter or shift register from 1 to 256 bits wide. LPM components can be used directly in a schematic file or as part of an HDL file. Refer to MAX+PLUS II Help or Altera's LPM Quick Reference (available on Altera's web site, www.altera.com) for information about specific LPM components.

Procedure

LPM Counter

Figure 54 shows a **gdf** representation of an 8-bit counter created from the Library of Parameterized Modules. The counter shown has an active-LOW asynchronous clear function (**aclr**) connected to a pin called **reset** and an active-HIGH asynchronous parallel load function (**aload**) connected to a pin called **load**.

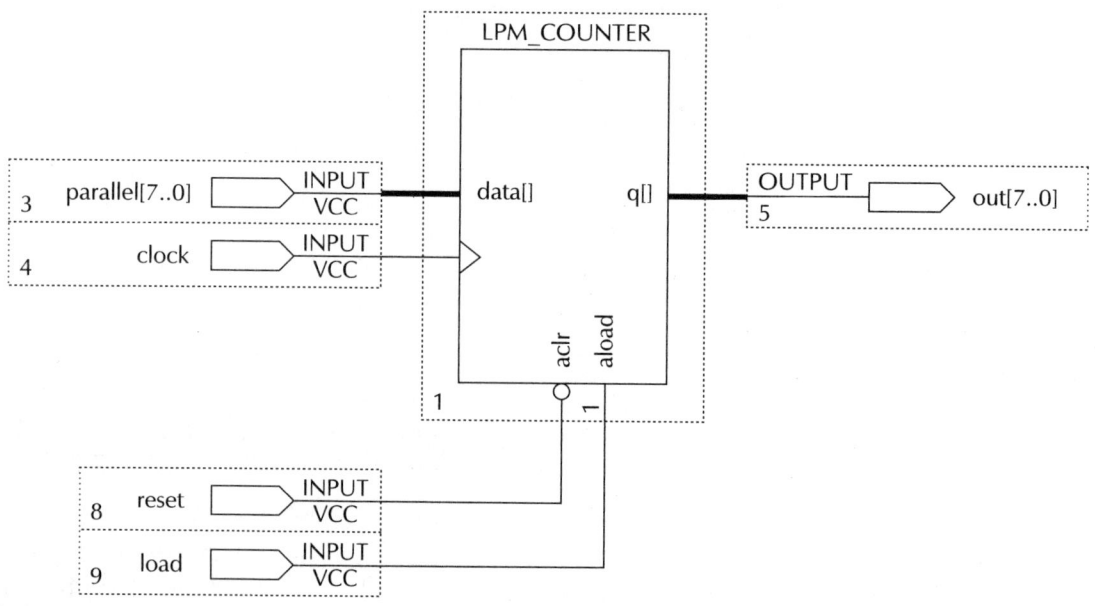

Figure 54   LPM Counter with Asynchronous Reset and Load

The eight outputs Q7 down to Q0 are shown as a single thick line, referred to as a **bus** (i.e., a collection of related signals). The eight outputs are designated by the label **out[7..0]**. When clock pulses are applied, the output counts from 0000 0000 to 1111 1111 (00H to FFH) and repeats.

The **aclr** input, shown as active-LOW, makes all **q[ ]** outputs LOW as soon as a LOW level is applied to the **reset** input. When a HIGH level is applied to the **load** pin, the **aload** input sets **q[ ]** outputs to the binary values at the **data[ ]** inputs (the eight pins **parallel[7..0]**).

An LPM module is specified by **ports** and **parameters**. A **port** is an input or output of the device, with a function such as clock, asynchronous clear, or asynchronous load. A **parameter** is a property of the block, such as **LPM_WIDTH**, which specifies how many bits its parallel input or output has. Some ports and parameters, such as **clock** and **LPM_WIDTH,** must be used in all instances of **lpm_counter.** Others, such as **aclr** and **LPM_DIRECTION**, are optional.

To enter an LPM component into a Graphic Design File, use the **Enter Symbol** dialog box, as shown in Figure 55.

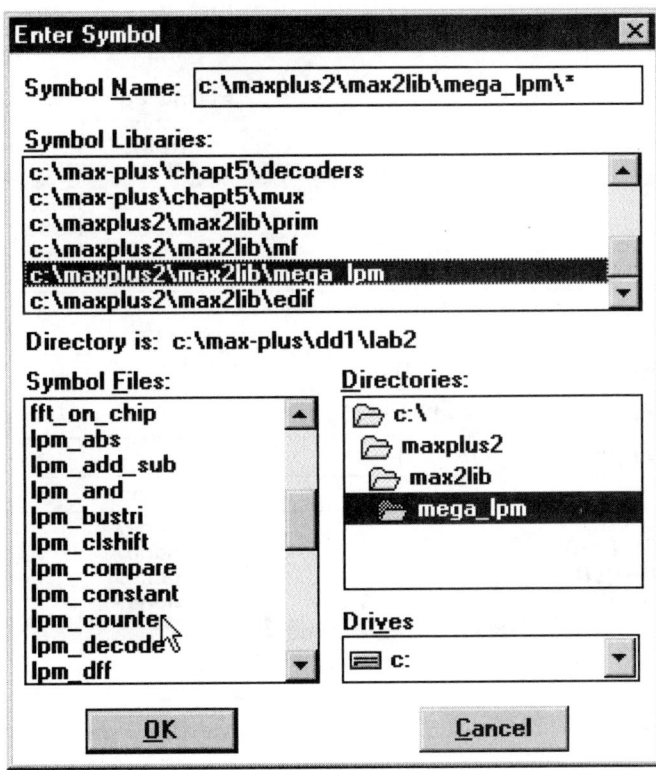

Figure 55   Dialog Box for lpm_counter Symbol

You can modify the port assignments by changing the **Used** or **Unused** option for each port listed in the **Port Status** portion of the dialog box, as shown in Figure 56. To make a port active-LOW, select the port name and select **all** in the **Inversion** box.

The parameters can be modified, as well. To modify a parameter, select its name in the **Parameters** box, type the new value in the **Parameter Value** box and click **Change**.

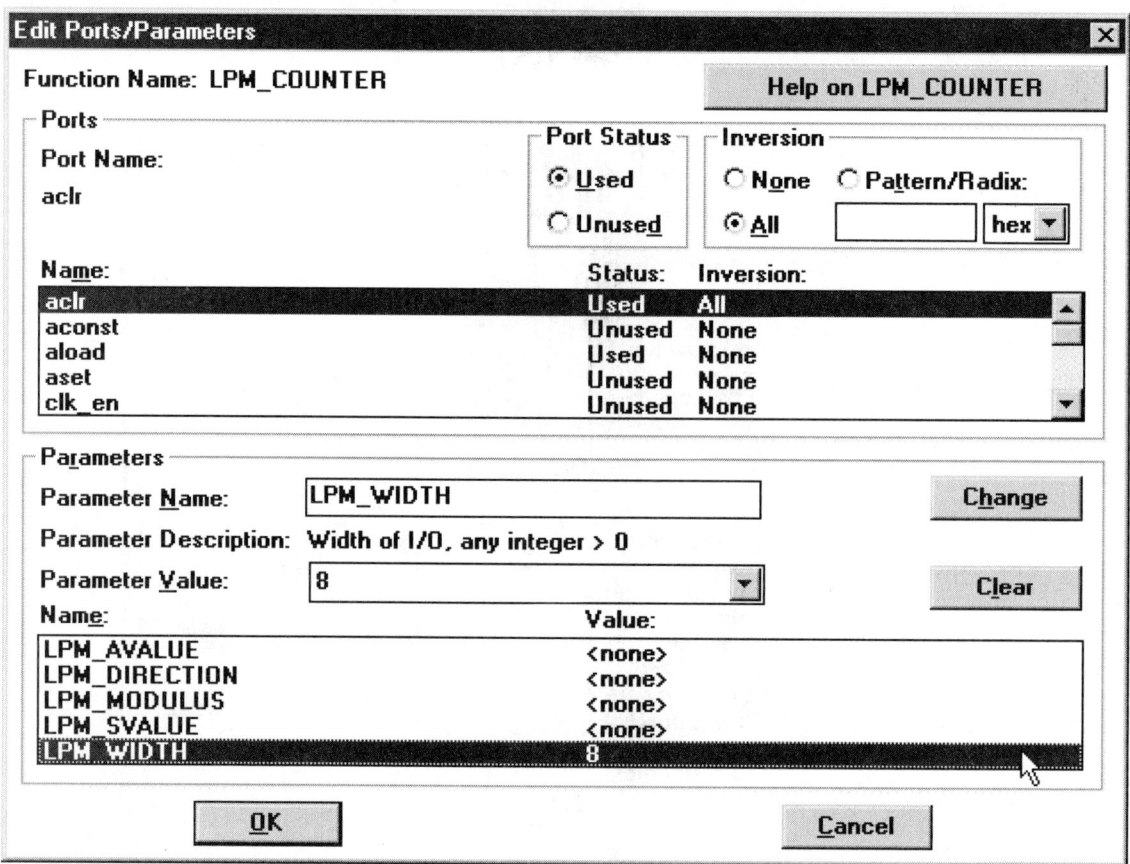

Figure 56   LPM Ports and Parameters Dialog Box

Creating an LPM Counter

1. Enter the LPM counter shown in Figure 54 in a new MAX+PLUS II file called **ct8lpm.gdf**. Use the ports and parameters shown. Save and compile the design. (Remember to **Set Project to Current File** first.)

2. Create a simulation file for the 8-bit LPM counter. The simulation should test the complete count sequence of the counter, the reset function, and the parallel load function. Show the simulation waveforms to your instructor.

Instructor's Initials: _____

3. Modify the LPM counter circuit to:

   a. change **aload** to an active-LOW signal;

   b. invert the outputs so that the on-board LEDs will count correctly; and

   c. include a switch debouncer, as shown in Figure 57.

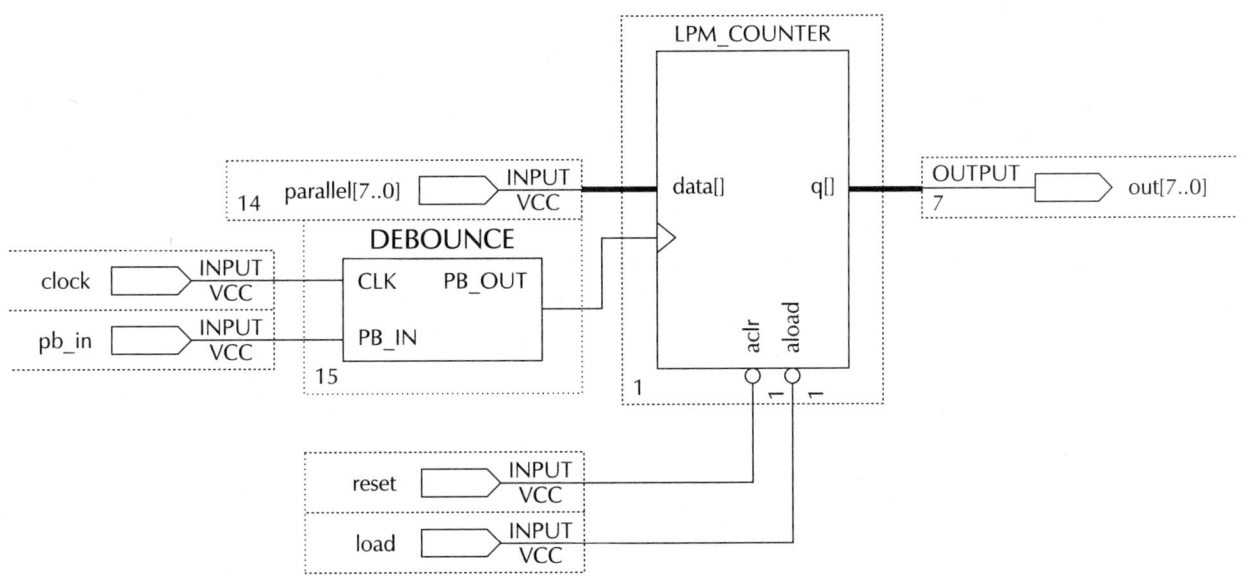

Figure 57   LPM Counter with Switch Debouncer

The port can be changed by right-clicking the LPM symbol and selecting **Edit Ports and Parameters** from the pop-up menu. Select the **aload** port in the dialog box and use the **Inversion** box to change it. Also invert the **q[ ]** port.

The clock signal from an on-board oscillator runs at 25.175 MHz, which is much too fast to use as a clock if we want to observe the counter output visually. We can use the module **debounce** to let us clock the counter manually. Copy the file **debounce.vhd** from the accompanying CD to your working directory, open the file in the MAX+PLUS II text editor, **Set Project to Current File**, and **Create Default Symbol**. Go back to **ct8lpm.gdf**, **Set Project to Current File**, and add the debouncer.

4. Refer to the table at the end of this lab (**EPM7128SLC84-7 Pin Assignments**). Assign **clock** to Pin 83 and **pb_in** (the manual clock pushbutton) to Pin 51. Assign **reset** to **PB2** (Pin 52 in the table) and **load** to DIP switch **SW2-8** (Pin 33). The parallel inputs should individually be assigned (parallel0, parallel1,...,parallel7) to the pins for the DIP switches SW1-1 to SW1-8. The counter outputs, out0, out1,...,out7, should be assigned to the pins corresponding to LED1 to LED8.

5. Wire the **reset, load, pb_in, parallel,** and **out** pins from the switches and LEDs to the connectors around the EPM7128S chip. Refer to the UP-1 User's Guide for pin numbers on the surrounding connectors. The **clock** pin (83) is hardwired to the chip and therefore does not need to be connected.

6. Download the LPM counter file to the Altera UP-1 board and demonstrate its operation to your instructor.

Modified LPM Counter

1. Refer to Altera's *LPM Quick Reference* or MAX+PLUS II Help to find the functions of the following ports and parameters for **lpm_counter: LPM_MODULUS, LPM_DIRECTION, LPM_SVALUE, updown, sset,** and **eq[ ]**.

2. Remove the **aload** and **data[ ]** ports from the LPM counter in Figure 54. Further modify the counter to perform the following functions:

   a. count up or down, depending on the state of an input called **direction;**

   b. synchronously set the counter output value to H"55" with an active-LOW input called **setn;**

   c. generate LOW-going pulses on two outputs: one for a count of 0 and one for a count of 7;

   d. set the counter modulus to 20 (decimal).

3. Create a simulation file that demonstrates these functions. (Do not use the debouncer when compiling your design just before you create a simulation. If you do, you will need 20×2^{16} pulses to go through the count once.) Show the simulation waveforms to your instructor.

   Instructor's Initials: _____

4. Add a debouncer to your design (as in Figure 57). Remember to invert your counter outputs so that the LEDs on the Altera UP-1 board will sequence correctly. Assign pins as needed. Compile the design and download it to the CPLD on the Altera UP-1 board. Demonstrate the operation of the counter to your instructor.

   Instructor's Initials: _____

LPM Shift Register

1. Refer to Altera's *LPM Quick Reference* or MAX+PLUS II Help to find the functions of the LPM shift register, **lpm_shiftreg**. Create a Graphic Design File for an 8-bit shift register having the following features:

   a. parallel load;

   b. synchronous clear;

   c. serial input, set to 0; and

   d. parallel outputs.

2. Create a simulation file for the above shift register that fully demonstrates the functions listed above. Show the simulation waveforms to your instructor.

   Instructor's Initials: _____

3. Add a clock divider or switch debouncer to the LPM shift register circuit. Assign pin numbers, compile the file, and download it to the CPLD on the Altera UP-1 board. Demonstrate the operation of the shift register to your instructor.

   Instructor's Initials: _____

Assignment Questions

1. Refer to MAX+PLUS II Help or Altera's *LPM Quick Reference*. State which ports and parameters are required for each of the following devices. Give the parameter values, if applicable.

   a. 32-bit counter with asynchronous clear and load

   b. 48-bit shift register with serial input, parallel output, synchronous clear

   c. 12-bit down-counter with an output that goes HIGH on an output count of 8

   d. a 10-bit shift register with serial input and output whose internal value can be synchronously set to H"3C0"

EPM7128LC84-7 Pin Assignments
Altera UP-1 Board

SEVEN SEGMENT DIGITS			
Function	**Pin**	**Function**	**Pin**
a1	58	a2	69
b1	60	b2	70
c1	61	c2	73
d1	63	d2	74
e1	64	e2	76
f1	65	f2	75
g1	67	g2	77
dp1	68	dp2	79

PUSHBUTTONS			
Function	**Pin**	**Function**	**Pin**
PB1	51	PB2	52

DIP SWITCHES			
Function	**Pin**	**Function**	**Pin**
SW1-1	12	SW2-1	15
SW1-2	16	SW2-2	17
SW1-3	18	SW2-3	21
SW1-4	20	SW2-4	25
SW1-5	22	SW2-5	27
SW1-6	24	SW2-6	29
SW1-7	28	SW2-7	31
SW1-8	30	SW2-8	33

LED OUTPUTS			
Function	**Pin**	**Function**	**Pin**
LED1	4	LED9	81
LED2	6	LED10	5
LED3	8	LED11	9
LED4	10	LED12	11
LED5	56	LED13	50
LED6	57	LED14	48
LED7	54	LED15	46
LED8	55	LED16	44

Unassigned: Pins 34, 35, 36, 37, 39, 40, 41, 45, 49, 80

Time-Division Multiplexing
(Design Project)

Name_____   Class_____   Date_____

Objectives	Upon completion of this laboratory exercise, you should be able to:

- Design, program, and test a circuit that transmits and receives four 4-bit numbers in a multiplexed sequence along a single transmission path.

- Use VHDL to program individual circuit components used in the 4-channel multiplexer.

- Use the MAX+PLUS II simulation tool to test the individual components of the 4-channel multiplexer and the whole circuit.

Reference	Dueck, Robert K., *Tutorial 2—Introduction to VHDL*
	Altera, MAX+PLUS II reference manuals:
	    Getting Started
	    VHSIC Hardware Description Language (VHDL)
	Altera, *LPM Quick-Reference Guide*

Equipment Required	Altera UP-1 University Lab Package:
	           UP-1 Circuit Board
	           ByteBlaster Download Cable
	           MAX+PLUS II Student Edition Software
	AC Adapter, minimum output: 7 VDC, 250 mA DC
	#24 solid-core wire
	Wire strippers

Experimental Notes

Time-division multiplexing is a method of improving the efficiency of a transmission system by sharing one transmission path among many signals. For example, if we wish to transmit four 4-bit numbers over a transmission line, we can send them one after the other, as shown in Figure 58.

Each bit is assigned a **time slot** in a sequence. During that time, it has sole access to the transmission line. When its time elapses, the next bit is sent and so on in sequence, until the channel assignment returns to the original location. In Figure 58, we see the 4-bit word p0 transmitted, LSB first, followed by p1, p2 (not shown), then p3.

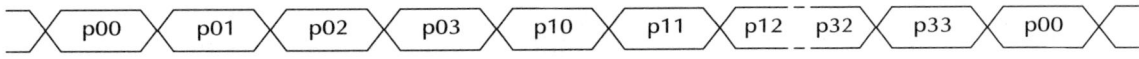

Figure 58   Time Division Multiplexing of Four 4-Bit Numbers

This can be achieved by connecting four 4-bit serial shift registers to the inputs of a 4-channel multiplexer, as shown in the diagram for the 4-channel transmitter in Figures 66 and 67 at the end of this lab.

When a MUX channel is selected by a counter, its corresponding shift register has sole access to the transmission path. This lasts for four clock pulses, after which access to the transmission path passes to the next channel. Channel selection is controlled by a counter and decoder.

At the receive end of the circuit, shown in Figures 68 and 69, the incoming data stream is directed to one of four serial shift registers. Since only one shift register accepts data at any time, this has the effect of demultiplexing the data.

After a shift register receives its data, the data are stored in a 4-bit latch. Transfer into the output latch happens only after the serial shifting has finished on that channel. During the shift period, old data are held in the latch so that the LED outputs do not flicker due to new data moving in.

Transmitter Implementation

The transmitter portion of this system consists of four main components: a multiplexer, four 4-bit shift registers, a 4-bit counter, and a decoder.

Multiplexer

Figure 59 shows the 4-channel multiplexer. The channel is selected by the upper two bits of a 4-bit counter. The data inputs are supplied by the serial outputs of the shift registers. The output, tx_out, provides the transmission path to the receiver portion of the circuit.

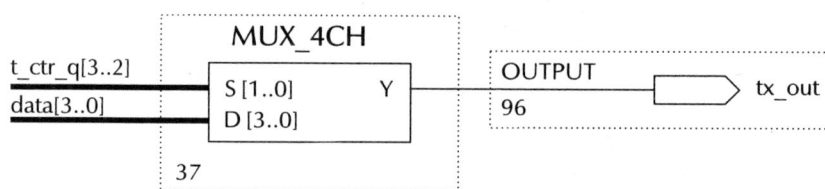

Figure 59   4-Channel Multiplexer

Shift Register

Figure 60 shows the symbol for one of the transmit shift registers. This 4-bit shift register has parallel inputs and a serial output. The serial output connects to one channel of the transmit multiplexer and sends four bits of data to the MUX when its channel is selected.

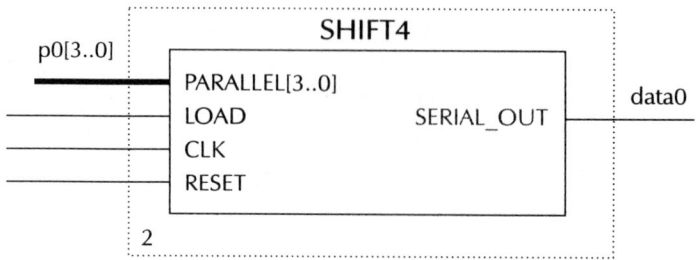

Figure 60   Transmit Shift Register

Four bits of data require four clock pulses, so each shift register is selected for four clock pulses.

When the shift register is not selected, its LOAD input is HIGH, causing parallel data from four DIP switches to be transferred into the shift register. The parallel data are shifted out next time the channel is selected.

The shift register also has an asynchronous reset.

Counter and Decoder

Figure 61 shows how the 4-bit counter and decoder control the sequence of selected channels in the transmitter circuit.

The decoder examines the two Most Significant Bits of the counter and generates a LOW value on the output that corresponds to the binary value of these two bits. For example, when the two counter MSBs are 00, then Y0 goes low and Y1, Y2, and Y3 are all HIGH.

The LOW on Y0 sets the Channel 0 shift register into serial shift mode and keeps it there for four clock pulses. At the same time the counter's two MSBs select Channel 0 on the multiplexer. (When channel 0 is selected, the counter goes through output values 0000, 0001, 0010, and 0011.) The transmitter serially shifts the four bits of the Channel 0 shift register to the transmission line output.

In a similar way, the next four counter states (0100, 0101, 0110, and 0111) select Channel 1 for serial transmission. Channel 2 transmits on the next four states, and then Channel 3. Following this, the transmit channel cycle repeats.

Note that the decoder module has a **clock** and a **setn** input. This particular decoder was designed with latching outputs that are loaded on the negative edge of the clock. This is to eliminate glitches on the decoder outputs that occur immediately after the decoder inputs change states. The decoder will still work without this output latch.

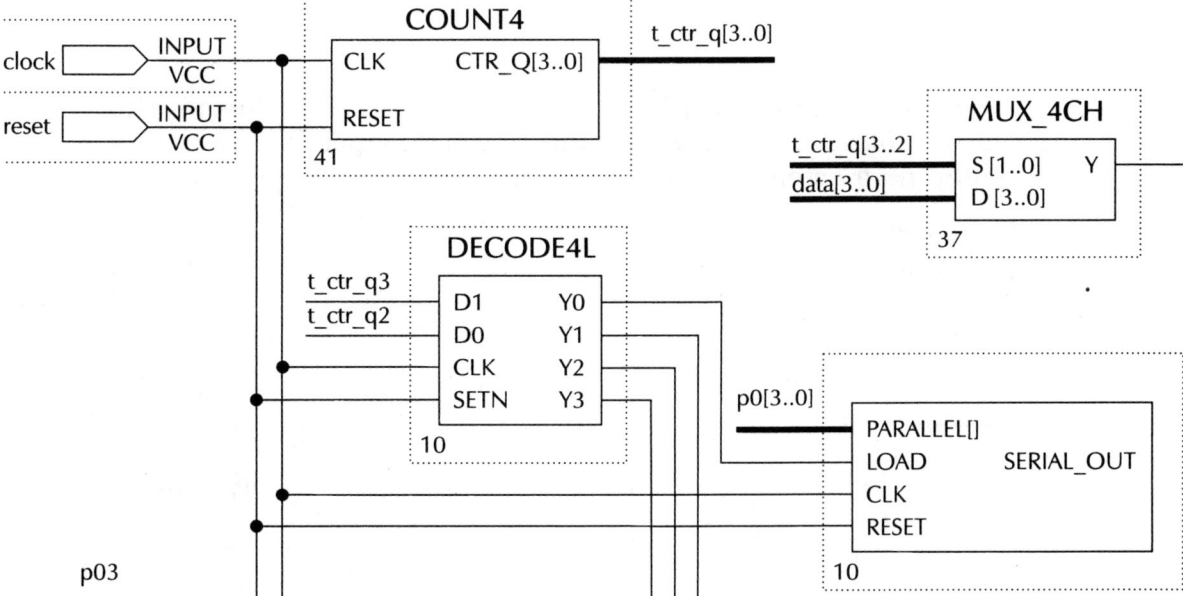

Figure 61 Counter and Decoder used as Channel Selector

Transmitter Design Requirements

1. Design each of the above modules and create a simulation for each one. At least two of the modules must be designed using VHDL. LPM modules are permissible. Again, note that the latching outputs of the decoder are optional.

2. As you are building up the circuit from components, think of some ways to determine if the assembled portion of the circuit is working. For example, create a simulation for the counter. If that works, connect a switch debouncer or clock divider to the counter, assign it some pins, and test if it produces a binary output on LEDs. Connect the counter outputs to the decoder and see if you can make the decoder pattern appear on LEDs, and so on.

3. Assign pins to the transmitter module as indicated in the table at the end of this handout. Parallel inputs correspond to the input DIP switches. Clock is Pin 83. Reset is PB2. Tx_out is Pin 4. (This is the pin for LED1. We will use this connection for testing only.)

4. Test the MUX transmitter by adding a divide-by-2^{23} clock divider. You should be able to see serial data appearing on LED1 of the Altera UP-1 board. You should be able to change the data pattern (lights blinking in a repeating sequence) by changing the pattern of DIP switches. Show this to your instructor.

Receiver Implementation

The Graphic Design File of a 4×4-bit receiver is shown in Figures 68 and 69 at the end of this lab. The 4-channel receiver consists of the following components: 4-bit counter, four 4-bit shift registers and latches, and a 4-bit decoder with complementary outputs.

Counter

As in the transmitter, the counter acts as a sequencing device. The upper two bits are decoded to select an active channel, each channel for four clock periods. The counter has an asynchronous reset.

Shift Register

Each shift register accepts serial input data whenever its ENABLE input is HIGH. At all other times, it holds its previous value, even though the input data are applied to its serial input. Shift register outputs are parallel (**q_out[3..0]**) as shown in Figure 62. The register

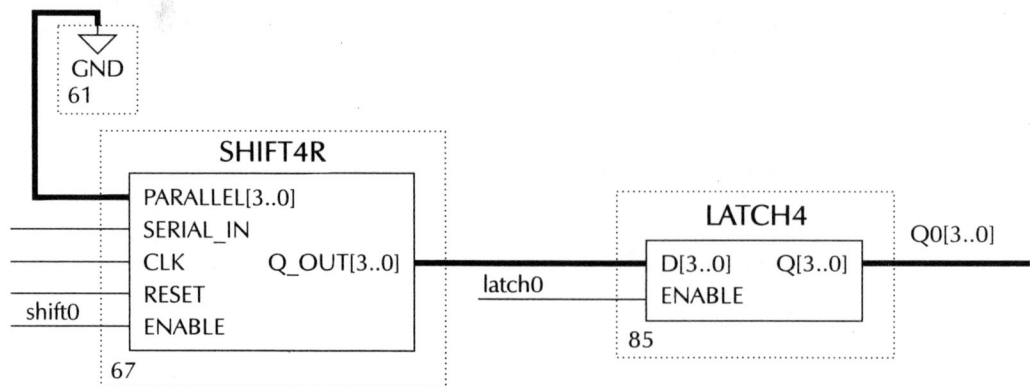

Figure 62   Receiver Shift Register and Latch

has an asynchronous reset. (Parallel inputs are shown grounded, but are not required at all in the receiver.)

Latch

The latch on each channel has four D inputs and four Q outputs. When ENABLE=1, Q follows D (transparent mode). When ENABLE=0, the latch holds its previous value (store mode). The latch is in store mode when the shift register is receiving serial data. This prevents the outputs from flickering while new data are being shifted in. When the shifting is complete, the latch becomes transparent.

Decoder

The decoder, shown in Figure 63, selects the active shift register and latch channels. When a channel is active, the corresponding decoder **shift** output goes HIGH and its **latch** output goes LOW. For example, when the counter selects channel 2 ($d_1 d_0 = 10$), **shift2** is HIGH and **latch2** is LOW.

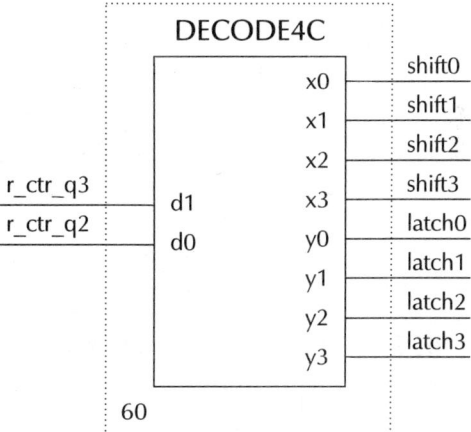

Figure 63 Decoder with Complementary Outputs

Receiver Design Requirements

1. Create each of the above components and create a simulation for each one. At least two of the components must be designed using VHDL. You may use LPM components if you so choose.

2. Add the receiver components to the drawing for the transmitter. Connect the receiver modules together and think of ways to simulate and/or test the partially constructed system.

3. Assign pin numbers to the receiver, according to the pin number assignment table at the end of the handout. Latch outputs are LEDs 1–16, Rx_in is Pin 37. **Clock** and **reset** lines should be connected to the transmitter **clock** and **reset** lines. **Change the output of the transmitter to Pin 35.** Compile and download the combined transmitter and receiver.

4. Connect a wire jumper from Pin 35 (tx_out) to Pin 37 (rx_in). The LED outputs should follow the changes in the DIP switch inputs. Demonstrate the operation of the circuit to your instructor.

Instructor's Initials: _____

Communication between Two Boards

Set up two boards so that the switches on one board change the LEDs on the other via a serial channel. Since communication will be possible in both directions, the connection is **full duplex**.

In order to communicate between two boards, it is necessary to designate one board as the **Master** and one as the **Slave**, as shown in Figure 64. The Master provides the clock to both boards and controls the reset function.

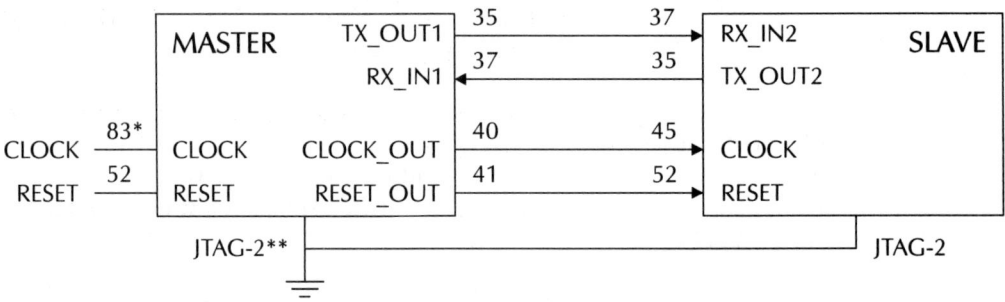

*Pin 83 (CLOCK) is connected internally. No external wire needed.

**This connection provides a common ground between the two boards. Connect pin 2 of the JTAG OUT socket to the same point on the other board. No external ground connection is required.

Figure 64   Communication Between Two Boards

You will need to make some small changes to your pin assignments to make this configuration work. These changes are as follows.

Master

1. Since the Master must supply **clock** and **reset** functions to the Slave, you must add two "pass-through" pins for this function, as shown in Figure 65. Assign these pass-through pins to Pins 40 (clock_out) and 41 (reset_out). Recompile and program the Master board.

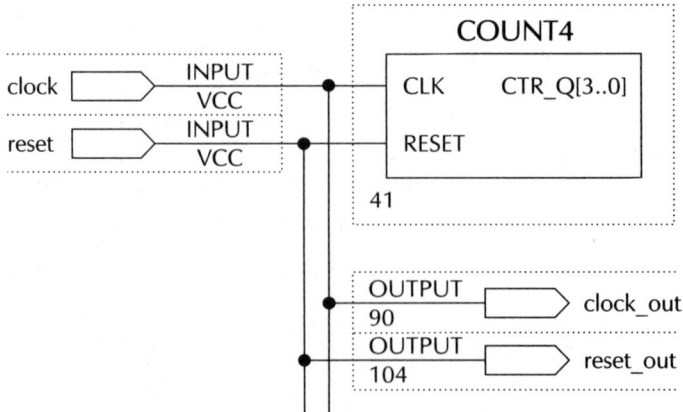

Figure 65   Clock and Reset Pass-Through (Master)

Slave

1. Change the clock pin assignment for the Slave clock to Pin 45. In the **Assign** menu, choose **Global Project Logic Synthesis**. Make sure that **Clock** and **Clear** are unchecked in the **Automatic Global** box.

2. Recompile and program the Slave board.

3. Connect the two boards as shown in Figure 64. Note that the Master clock is hardwired from the on-board oscillator. The ground pins (Pin 2 on the JTAG OUT connector) are hardwired to grounds on each board. However, the ground pins must be connected between boards to ensure a common reference.

4. The switches on either board should control the LEDs on the other. Demonstrate this function to your instructor.

Instructor's Initials: _____

EPM7128LC84-7 Pin Assignments
Altera UP-1 Board

SEVEN SEGMENT DIGITS			
Function	Pin	Function	Pin
a1	58	a2	69
b1	60	b2	70
c1	61	c2	73
d1	63	d2	74
e1	64	e2	76
f1	65	f2	75
g1	67	g2	77
dp1	68	dp2	79

PUSHBUTTONS			
Function	Pin	Function	Pin
PB1	51	PB2	52

DIP SWITCHES			
Function	Pin	Function	Pin
SW1-1	12	SW2-1	15
SW1-2	16	SW2-2	17
SW1-3	18	SW2-3	21
SW1-4	20	SW2-4	25
SW1-5	22	SW2-5	27
SW1-6	24	SW2-6	29
SW1-7	28	SW2-7	31
SW1-8	30	SW2-8	33

EPM7128LC84-7 Pin Assignments
Altera UP-1 Board (*continued*)

LED OUTPUTS			
Function	**Pin**	**Function**	**Pin**
LED1	4	LED9	81
LED2	6	LED10	5
LED3	8	LED11	9
LED4	10	LED12	11
LED5	56	LED13	50
LED6	57	LED14	48
LED7	54	LED15	46
LED8	55	LED16	44

Unassigned: Pins 34, 35, 36, 37, 39, 40, 41, 45, 49, 80

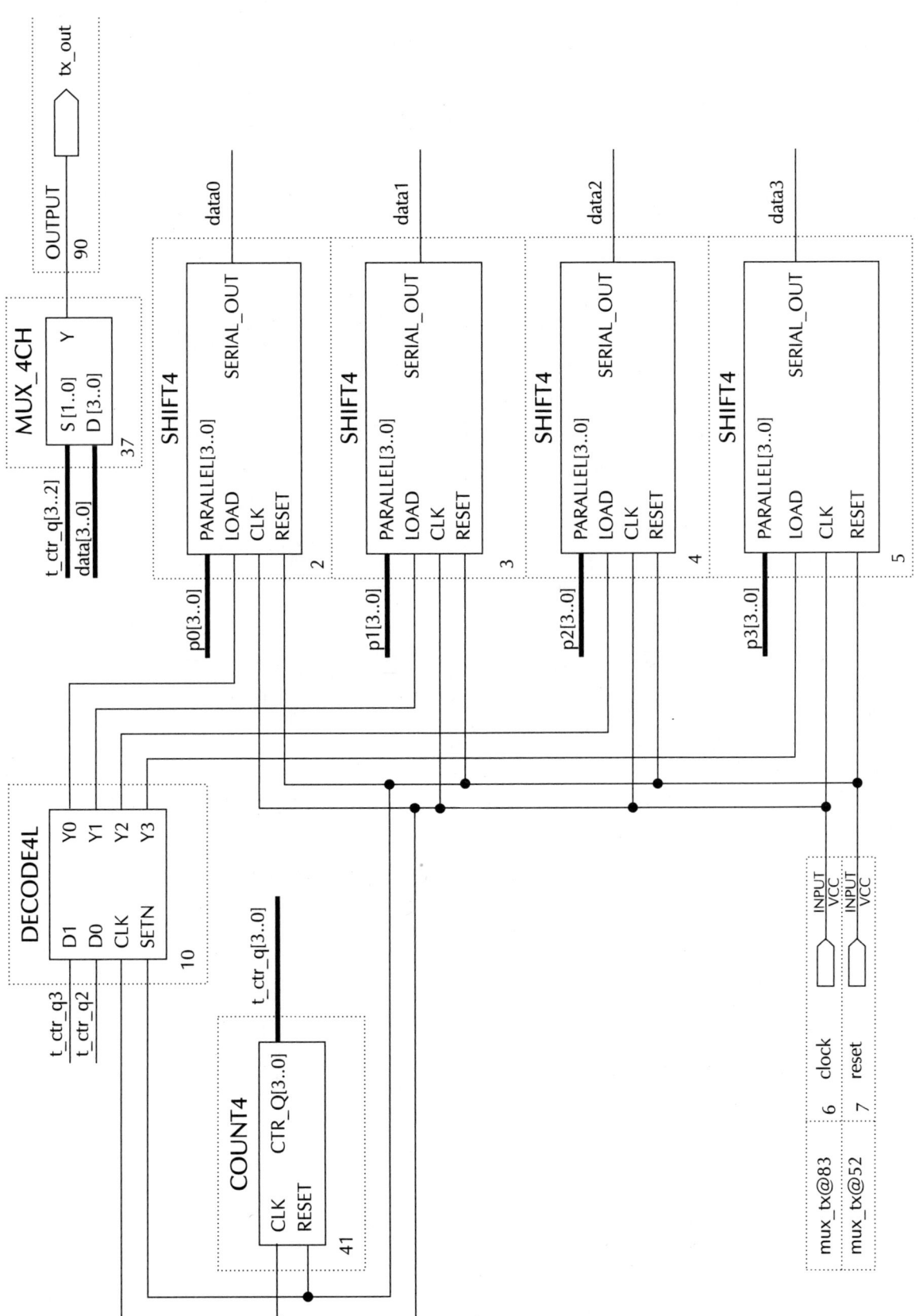

Figure 66   Multiplexing Transmitter

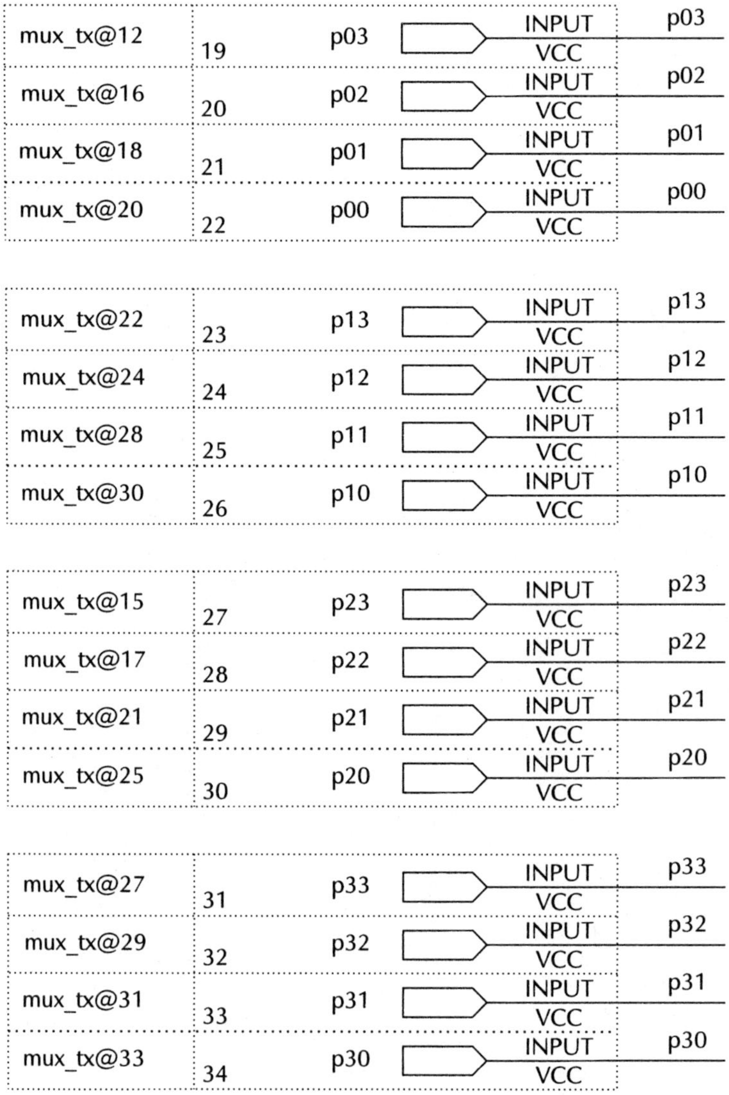

mux_tx@12	19	p03	INPUT / VCC	p03
mux_tx@16	20	p02	INPUT / VCC	p02
mux_tx@18	21	p01	INPUT / VCC	p01
mux_tx@20	22	p00	INPUT / VCC	p00

mux_tx@22	23	p13	INPUT / VCC	p13
mux_tx@24	24	p12	INPUT / VCC	p12
mux_tx@28	25	p11	INPUT / VCC	p11
mux_tx@30	26	p10	INPUT / VCC	p10

mux_tx@15	27	p23	INPUT / VCC	p23
mux_tx@17	28	p22	INPUT / VCC	p22
mux_tx@21	29	p21	INPUT / VCC	p21
mux_tx@25	30	p20	INPUT / VCC	p20

mux_tx@27	31	p33	INPUT / VCC	p33
mux_tx@29	32	p32	INPUT / VCC	p32
mux_tx@31	33	p31	INPUT / VCC	p31
mux_tx@33	34	p30	INPUT / VCC	p30

Figure 67   Inputs for Multiplexing Transmitter

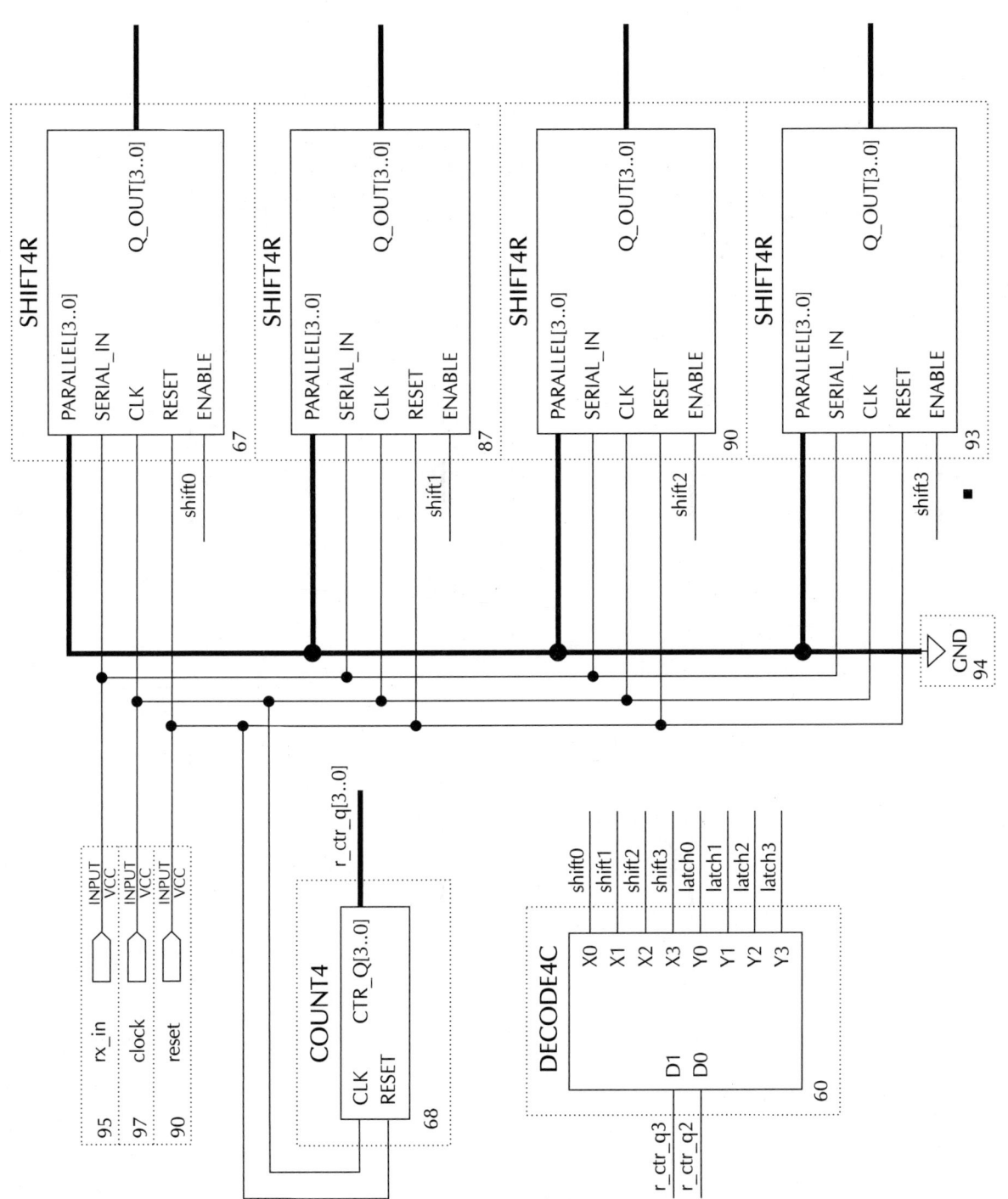

Figure 68   Demultiplexing Receiver (1 of 2)

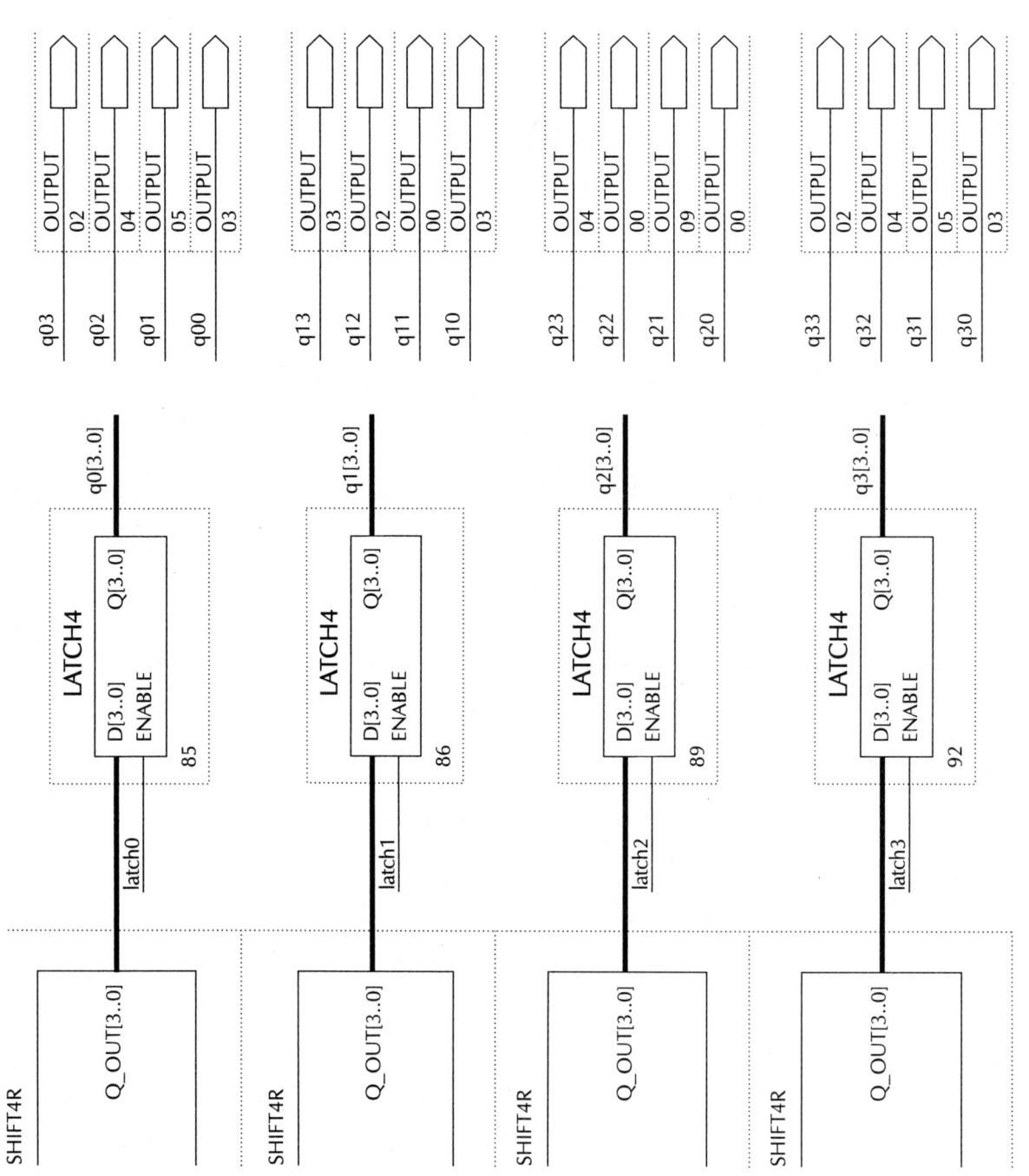

Figure 69   Demultiplexing Receiver (2 of 2)

DAC Function Generator

Name_____   Class_____   Date _____

Objectives   Upon completion of this laboratory exercise, you should be able to:

- Interface an integrated circuit Digital-to-Analog Converter to an EPM7128S CPLD on an Altera UP-1 circuit board.

- Program an Altera CPLD to control the DAC for manual input.

- Program an Altera CPLD to make the DAC generate sawtooth, square, and triangle waves, with function selectable by DIP switch input.

Reference   Dueck, Robert K., *Fundamentals of Digital Electronics* (© 1994, West Publishing) (pp. 104–112 of this lab manual)
Section 16.2, Digital-to-Analog Conversion (pp. 707–16)
Section 17.3, EPROM Application: Digital Function Generator (pp.758–61)

Equipment Required   Altera   UP-1 University Lab Package:
                    UP-1 Circuit Board
                    ByteBlaster Download Cable
                    MAX+PLUS II Student Edition Software
AC Adapter, minimum output: 7 VDC, 250 mA DC
Solderless Breadboard
±12 V Power Supply
DAC0808 Digital-to-Analog Converter
High-speed op-amp
0.1-µF capacitor
0.75-pF capacitor
2.7-kΩ resistor
4.7-kΩ resistor
6.8-kΩ resistor
5-kΩ potentiometers (2)
#24 solid-core wire
Wire strippers
Anti-static wrist strap

DAC/CPLD Interface

1. Wire the circuit for a DAC0808 digital-to-analog converter onto a solderless breadboard, as shown in Figure 71 and explained in Step 2. (Note that the DAC0808 is a direct replacement for the MC1408.)

2. **DAC Digital Inputs:** Create a MAX+PLUS II file that will read the values of 8 DIP switches at a set of CPLD input pins and pass them through the CPLD to a set of output pins without modification (i.e., an internal pin-to-pin connection). We can keep these same input and output pin connections for later parts of the lab, where other functions are used.

   Suggested pin assignments   Use the assignments shown in the table at the end of this lab for SW1-1 to SW1-8 (DIP switch inputs). Disconnect any LED connections for LED1 through LED8. Use the pins normally assigned to these LEDs to connect from the CPLD to the DAC inputs.

   Connect the DIP switches to the prototyping headers for the EPM7128S chip on the Altera UP-1 board. Connect the CPLD outputs to the DAC inputs. Compile and download the file to the Altera UP-1 board.

3. The resistor networks shown in Figure 70 allow us to set our input reference current and output gain to values within a specified range. Using the values shown in Figure 70, fill in Table 9 for the cases when V_o is at minimum and maximum, and when the pots are at their midpoint values. (R_F is the resistance of the op-amp feedback network.) Assume the DAC input is set to 1111 1111. Show calculations in the space provided on the next page.

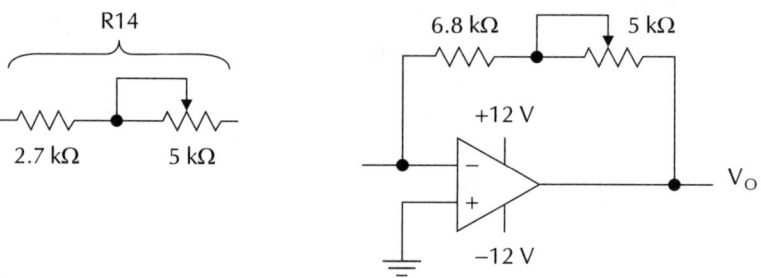

Figure 70   Resistor Networks for DAC0808/MC1408

Table 9   DAC Output Range

	$R_{14}(\Omega)$	$R_F(\Omega)$	$I_{ref}(mA)$	$I_o(mA)$	$V_o(V)$
Minimum V_o					
Maximum V_o					
Pots at midpoint					

Calculations:

4. Calibrate the DAC as follows:

   a. Connect a Digital Multimeter (DMM) to the op-amp output. Apply power to the circuit and adjust the feedback pot for the minimum value of voltage. Set the input code to 10000000.

   b. Adjust the R_{14} pot so that the output voltage of the op-amp is 3.4 volts. What value of I_{ref} does this correspond to? I_{ref} = _____.

   c. Set the feedback pot so that the output voltage is 5 volts. What value of R_F does this correspond to? R_F = _____.

5. Measure the output voltage of the DAC circuit for the following digital input values:

INPUT CODE	OUTPUT VOLTAGE (CALCULATED)	OUTPUT VOLTAGE (MEASURED)	ERROR (VOLTS)	ERROR (LSB)
00000000				
00000001				
00000011				
00000111				
00001111				
10000000				
11000000				
11100000				
11111111				

6. State the calculated value of resolution (i.e., the value of 1 LSB) for this DAC. From the above table, state the offset error, gain error, and linearity error in LSB.

   Resolution   = _____ volts

   Offset Error   = _____ LSB

   Gain Error   = _____ LSB

   Linearity Error = _____ LSB

Instructor's Initials: _____

DAC Ramp Generator

1. Refer to the DAC-based ramp generator under the heading "Sawtooth Waveform Generator" from *Fundamentals of Digital Electronics* or refer directly to page 105 of this manual. Create a similar circuit to the ramp generator in Figure 73 by programming an 8-bit counter into the EPM7128S CPLD on the Altera UP-1 board. The DAC interface in Figure 71 should not be changed. The counter driving the ramp generator should be clocked at about 1.6 MHz.
(The on-board oscillator runs at 25.175 MHz.)

2. Assign pins to the counter design file so that the counter outputs are the same as the pins connected to the DAC inputs. Compile and download the MAX+PLUS II counter file.

3. Connect an oscilloscope to the DAC op-amp output. Draw the sawtooth waveform generated by the DAC. Measure its period and calculate the sawtooth frequency.

   T= _____; f = _____.

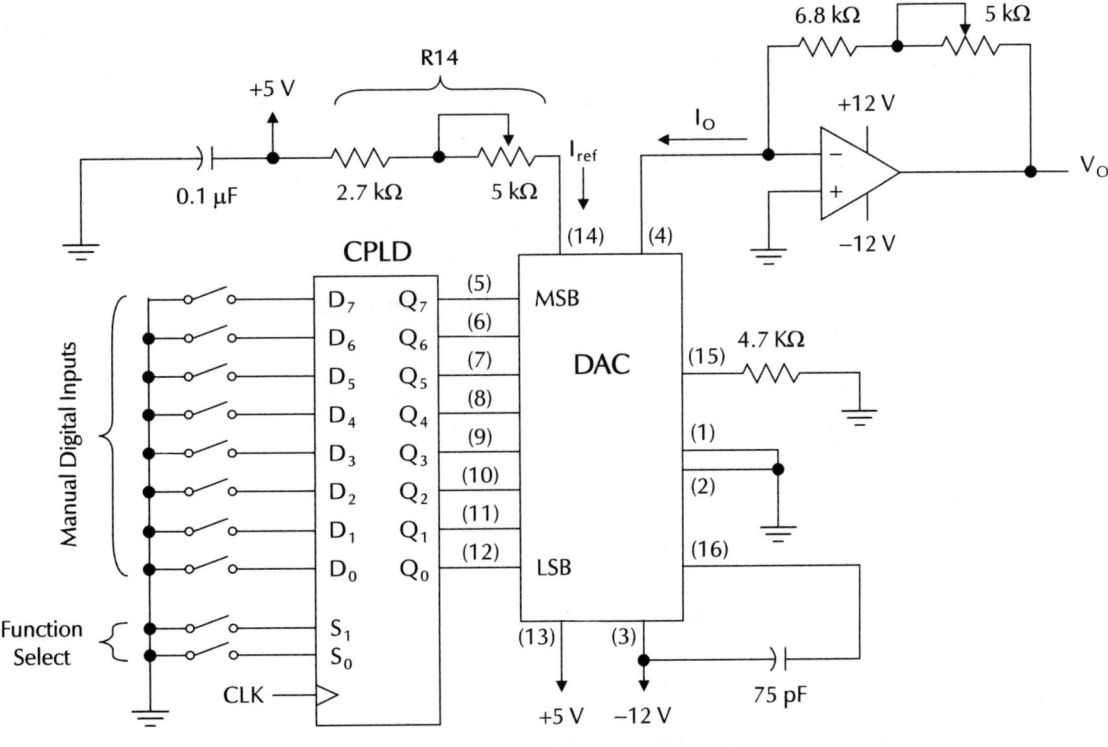

Figure 71   DAC Function Generator

Divide the counter clock frequency (f_c) by the DAC output frequency (f_{DAC}) to get an estimate of the number of clock pulses per sawtooth cycle. Compare this to the ideal value and determine the % error. Also state the % error of an oscilloscope measurement.

f_c/f_{DAC} = _____ clock cycles (measured)

f_c/f_{DAC} = _____ clock cycles (ideal)

%error (frequency measurement)   = _____

%error (oscilloscope screen)   = _____

Instructor's Initials: _____

Other Functions

1. Program the EPM7128S CPLD in Figure 71 to incorporate **sawtooth, triangle,** and **square** waveforms, as well as a **pass-through for manual DIP switch input** to the DAC. Program the CPLD using VHDL. (Hint: The function driving the Q outputs will depend on the binary value of S_1S_0.)

   Consider the following:

WAVEFORM	CONDITION
Sawtooth	Binary-increasing input 00 to FF. Input then drops to 00 (can be generated by counter function applied to input)
Square	Input = 00 for first half-cycle; = FF for second half-cycle
Triangle	Input increases linearly to positive peak, then decreases linearly to negative peak (no sharp drop-off).

   Use the following input switch combinations for function select:

S_1	S_0	FUNCTION
0	0	Manual pass-through
0	1	Sawtooth
1	0	Square
1	1	Triangle

2. Connect an oscilloscope to the op-amp output and demonstrate the operation of the waveform generator.

Instructor's Initials: _____

EPM7128LC84-7 Pin Assignments
Altera UP-1 Board

SEVEN SEGMENT DIGITS			
Function	**Pin**	**Function**	**Pin**
a1	58	a2	69
b1	60	b2	70
c1	61	c2	73
d1	63	d2	74
e1	64	e2	76
f1	65	f2	75
g1	67	g2	77
dp1	68	dp2	79

PUSHBUTTONS			
Function	**Pin**	**Function**	**Pin**
PB1	51	PB2	52

DIP SWITCHES			
Function	**Pin**	**Function**	**Pin**
SW1-1	12	SW2-1	15
SW1-2	16	SW2-2	17
SW1-3	18	SW2-3	21
SW1-4	20	SW2-4	25
SW1-5	22	SW2-5	27
SW1-6	24	SW2-6	29
SW1-7	28	SW2-7	31
SW1-8	30	SW2-8	33

LED OUTPUTS			
Function	**Pin**	**Function**	**Pin**
LED1	4	LED9	81
LED2	6	LED10	5
LED3	8	LED11	9
LED4	10	LED12	11
LED5	56	LED13	50
LED6	57	LED14	48
LED7	54	LED15	46
LED8	55	LED16	44

Unassigned: Pins 34, 35, 36, 37, 39, 40, 41, 45, 49, 80

Reference Material

The following reference material is extracted from: Dueck, Robert K. *Fundamentals of Digital Electronics*, (© 1994, West Publishing). This material contains examples of DAC functions similar to those in Lab 7, but implemented with standard TTL and CMOS components, rather than a CPLD.

Op Amp Buffering of MC1408

The MC1408 DAC will not drive much of a load on its own, particularly when the Range input is grounded. We can use an operational amplifier to increase the output voltage and current. This allows us to select the lower voltage range for faster switching while retaining the ability to drive a reasonable load. The output voltage is limited only by the op amp supply voltages. We use a 34071 High Slew Rate op amp for fast switching.

Figure 72 shows such a circuit. R_{14} adjusts the maximum output current of the MC1408 and therefore the maximum output voltage of the op amp. The 0.1-μF capacitor decouples the +5-V supply. (The manufacturer actually recommends that the +5-V logic supply not be used as a reference voltage. It doesn't matter for a demonstration circuit, but may introduce noise that is unacceptable in a commercial design.) The 75-pF capacitor is for phase compensation.

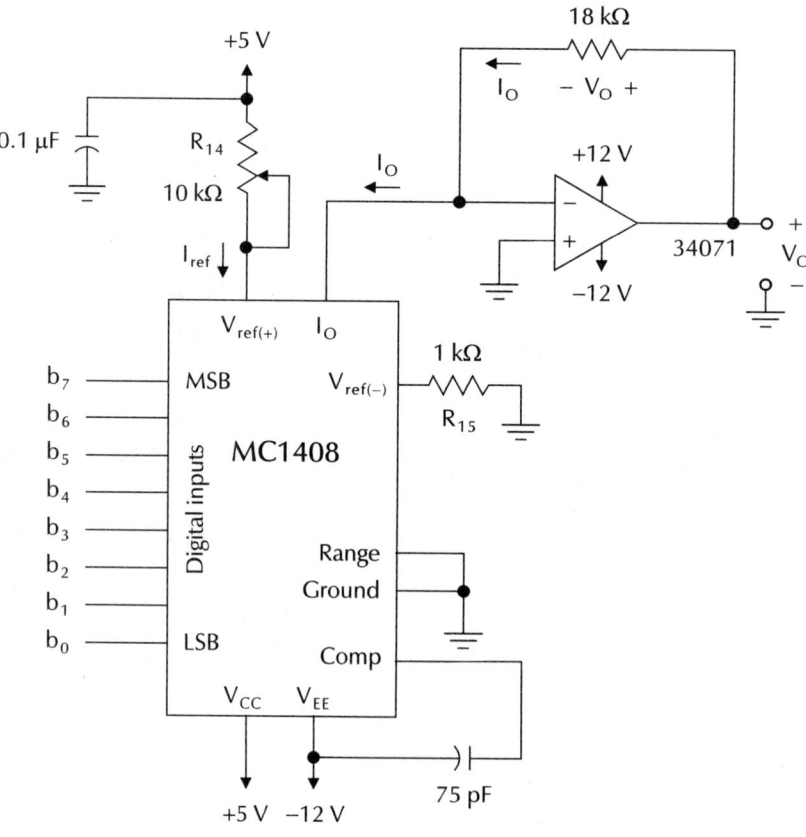

Figure 72 DAC with Op Amp Buffering

V_o is positive because the voltage drop across R_F is positive with respect to the virtual ground at the op amp (−) input. This feedback voltage is in parallel with (i.e., the same as) the output voltage, since both are measured from output to ground.

The output voltage is given by:

$$V_o = \left[\frac{b_7}{2} + \frac{b_6}{4} + \frac{b_5}{8} + \frac{b_4}{16} + \frac{b_3}{32} + \frac{b_2}{64} + \frac{b_1}{128} + \frac{b_0}{256}\right]\frac{R_F}{R_{14}}V_{ref}$$

V_o can, in theory, be any positive value less than the op amp positive supply ($+12$ V in this case). Any attempt to exceed this voltage makes the op amp saturate. The actual maximum value, if not the same as the op amp's saturation voltage, depends on the values of the two resistors.

Example Refer to the DAC circuit in Figure 72. To what value must R_{14} be set to make $V_o = +6$ V for $b_7b_6b_5b_4b_3b_2b_1b_0 = 10000000$? When $b_7b_6b_5b_4b_3b_2b_1b_0 = 11111111$ what is the value of V_o?

Solution To make $V_o = +6$ V, $I_o = V_o / R_F = 6$ V $/ 18$ k$\Omega = 333$ μA.

$$I_o = \tfrac{1}{2}(V_{ref}(+) / R_{14})$$
$$R_{14} = \tfrac{1}{2}(V_{ref}(+) / I_o = \tfrac{1}{2}(5 \text{ V} / 333 \text{ }\mu\text{A})$$
$$= 7.5 \text{ k}\Omega$$

For $b_7b_6b_5b_4b_3b_2b_1b_0 = 11111111$, V_o is:
$$V_o = (255 / 256)(18 \text{ k}\Omega / 7.5 \text{ k}\Omega)(5 \text{ V}) = 11.953 \text{ V}$$
$$(= \text{ full scale} - 1 \text{ LSB})$$

Sawtooth Waveform Generator

Example Figure 73 shows the circuit of an analog ramp (sawtooth) generator built from an MC1408 DAC, an op amp, and an 8-bit synchronous counter. (A ramp generator has numerous analog applications, such as sweep generation in an oscilloscope and frequency sweep in a spectrum analyzer.)

Briefly explain the operation of the circuit and sketch the output waveform. Calculate the step size between analog outputs resulting from adjacent codes. Assume that the DAC is set for $+6$-V output when the input code is 10000000.

Calculate the output sawtooth frequency when the clock is running at 1 MHz.

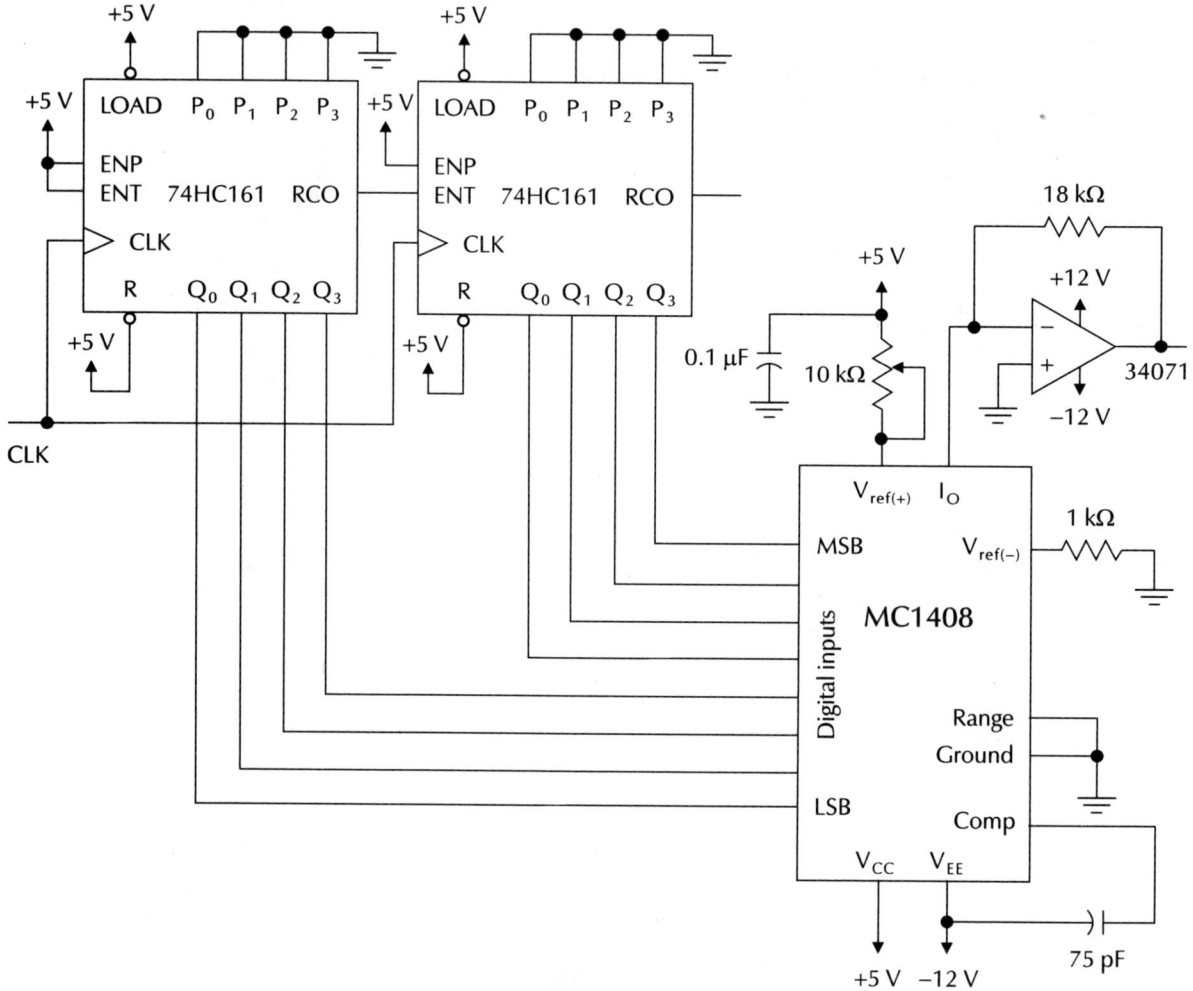

Figure 73   DAC Ramp Generator

Solution   The 8-bit counter cycles from 00000000 to 11111111 and repeats continuously. This is a total of 256 states.

The DAC output is 0 V for an input code of 00000000 and (12 V − 1 LSB) for a code of 11111111. We know this because a code of 10000000 always gives an output voltage of half the full-scale value (6 V = 12 V/2), and the maximum code gives an output that is one step less than the full-scale voltage. The step size is 12 V/256 steps = 46.9 mV/step. The DAC output advances linearly from 0 to (12 V − 1 LSB) in 256 clock cycles.

Figure 74 shows the analog output plotted against the number of input clock cycles. The ramp looks smooth at the scale shown. A section enlarged 32 times shows the analog steps resulting from eight clock pulses.

One complete cycle of the sawtooth waveform requires 256 clock pulses. Thus, if f_{CLK} = 1 MHz, f_o = 1 MHz/256 = 3.9 kHz.

(Note that if we do not use a high slew rate op amp, the sawtooth waveform will not have vertical sides.)

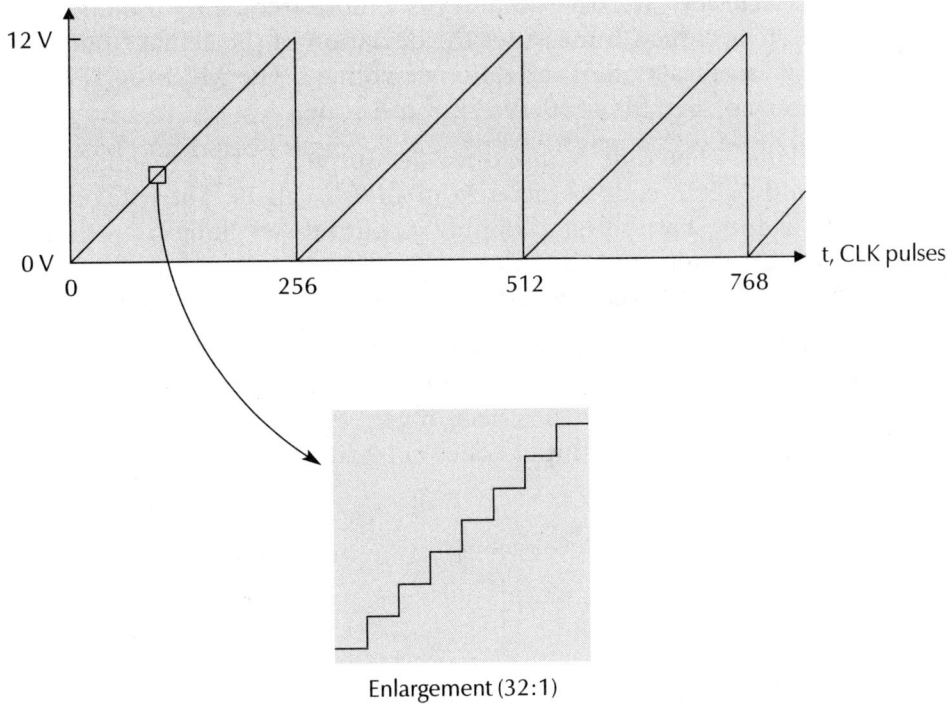

Enlargement (32:1)

Figure 74   Sawtooth Waveform Output of Circuit in Figure 16.14

DAC Performance Specifications

A number of factors affect the performance of a digital-to-analog converter. The major factors are briefly described below.

Monotonicity   The output of a DAC is monotonic if the magnitude of the output voltage increases every time the input code increases. Figure 75 shows the output of a DAC that increases monotonically and the output of a DAC that does not.

Absolute Accuracy   This is a measure of DAC output voltage with respect to its expected value.

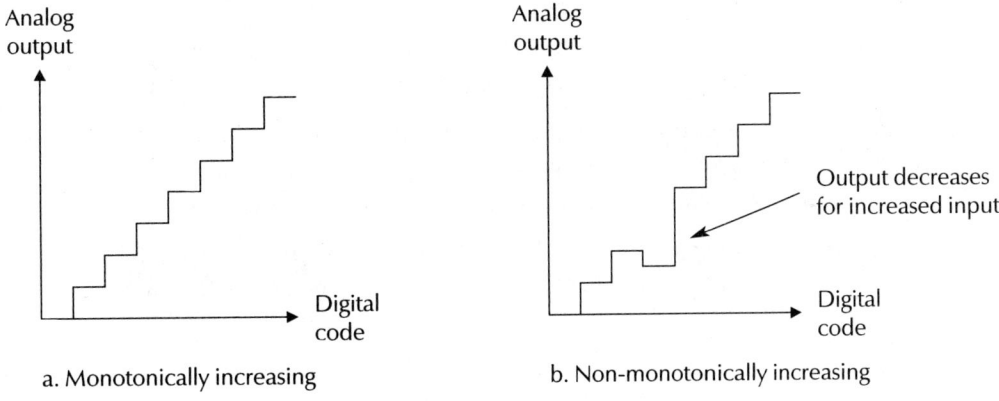

a. Monotonically increasing   b. Non-monotonically increasing

Figure 75   DAC Monotonicity

Relative Accuracy   Relative accuracy is a more frequently used measurement than absolute accuracy. It measures the deviation of the actual from the ideal output voltage as a fraction of the full-scale voltage. The MC1408 DAC has a relative accuracy of $\pm\frac{1}{2}$ LSB = $\pm0.195\%$ of full scale.

Settling Time   The time required for the outputs to switch and settle to within $\pm\frac{1}{2}$ LSB when the input code switches from all 0s to all 1s. The MC1408 has a settling time of 300 ns for 8-bit accuracy, limiting its output switching frequency to 1/300 ns = 3.33 MHz. Depending on the value of R_4, the output resistor, the settling time of the MC1408 may increase to as much as 1.2 μs when the Range input is open.

Gain Error   Gain error primarily affects the high end of the output voltage range. If the gain of a DAC is too high, the output saturates before reaching the maximum output code. Figure 76 shows the effect of gain error in a 3-bit DAC. In the high gain response, the last two input codes (110 and 111) produce the same output voltage.

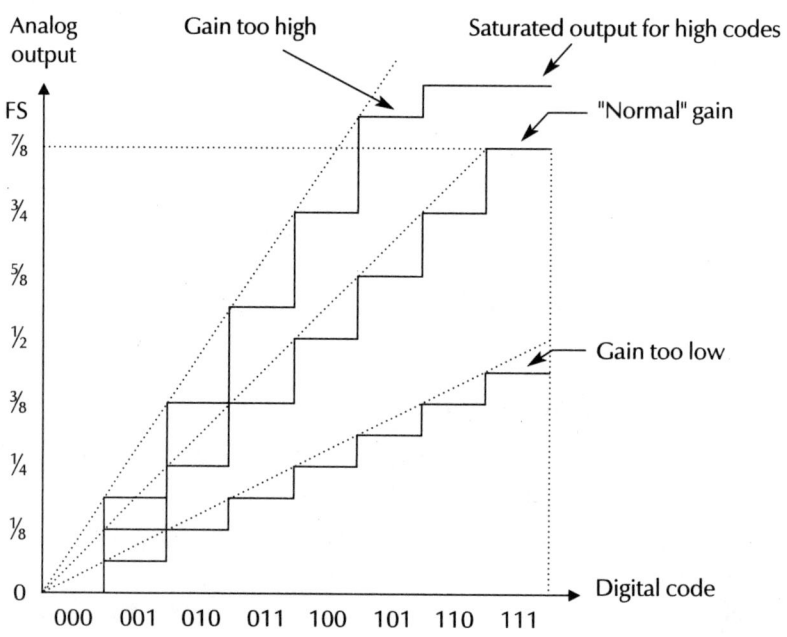

Figure 76   DAC Gain Errors

Linearity Error   This error is present when the analog output does not follow a straight-line increase with increasing digital input codes. Figure 77 shows this error.

Differential Nonlinearity   This specification measures the difference between actual and expected step size of a DAC when the input code is changed by 1 LSB. An actual step that is smaller than the expected step can result in a nonmonotonic output.

Offset Error   This error occurs when the analog output of a positive-value DAC is not 0 V when the input code is all 0s. Figure 78 shows the effect of offset error.

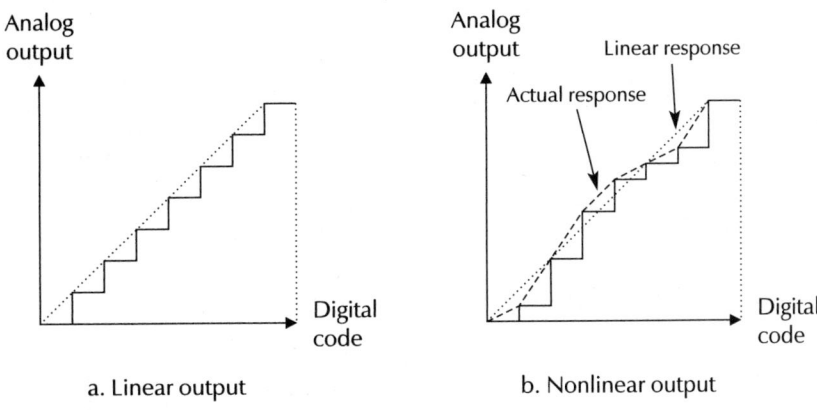

Figure 77   DAC Linearity

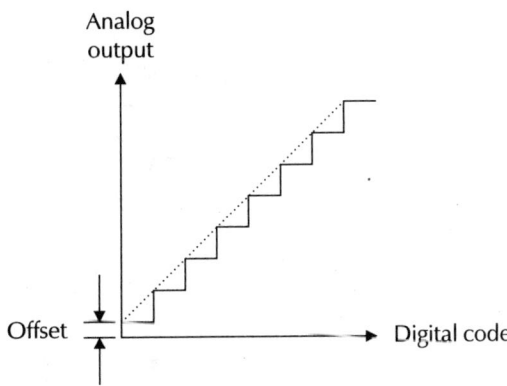

Figure 78   DAC Offset Error

EPROM Application: Digital Function Generator

An EPROM can be used as the central component of a digital function generator. Other components in the system include a clock generator, a counter, a digital-to-analog converter, and an output op amp buffer. The portion of the circuit including the last three of these components is shown in Figure 79.

The generator can produce the usual analog waveforms—sine, square, triangle, sawtooth—and any other waveforms that you wish to store in the EPROM. A single cycle of each waveform is stored as 256 consecutive 8-bit numbers. For example, the data for one cycle of the square wave data are stored at addresses 0100H to 01FFH, as shown in hexadecimal form in Table 10. Triangle and sawtooth waveform data are stored at addresses 0200H–02FFH and 0300H–03FFH, respectively.

The most significant bits of the EPROM address select the waveform function by selecting a block of 256 address. The 8 least significant bits of the EPROM address are connected to an 8-bit (mod-256) counter, which continuously cycles through the 256 selected addresses. A 2764 EPROM (8K × 8) has 13 address lines. After the eight lower

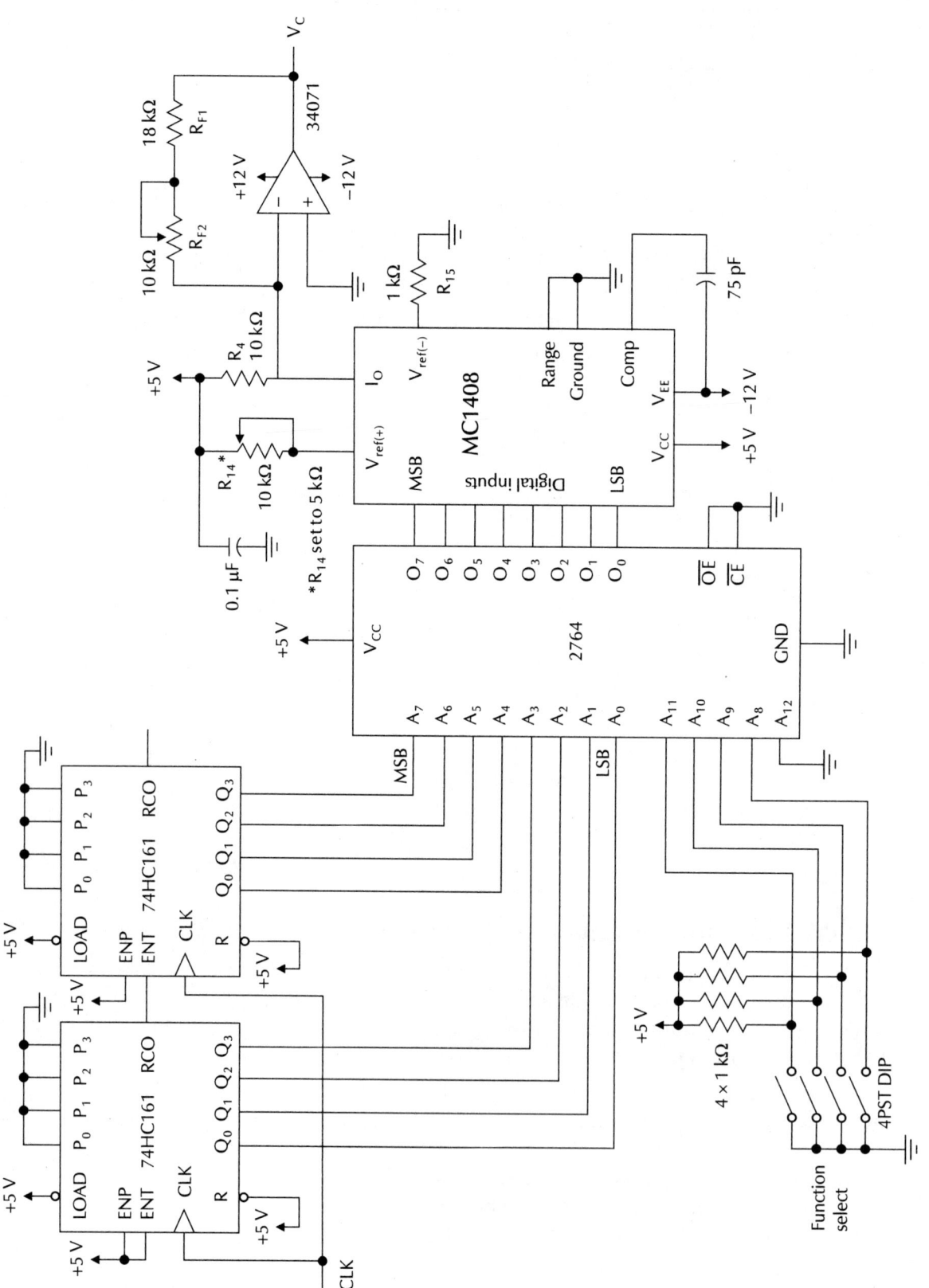

Figure 79   Digital Function Generator

lines are accounted for, the remaining five lines can be used to select up to 32 digital functions. With the four binary Function Select switches, we can potentially select 16 functions.

For example, to select the Triangle function, inputs $A_{11}...A_8$, which comprise the most significant digit of the EPROM address, are set to 0010. Thus, the 8-bit counter cycles through addresses 0200H–02FFH, the location of the triangle wave data. The Square Wave function is selected by setting $A_{11}...A_8$ to 0001, thus selecting the address block 0100–01FF. Other functions can be similarly selected.

The data at each address are sent to the D/A converter (MC1408), which, in combination with the op amp, is configured to produce a bipolar (both positive and negative) output. (We use a 34071 high slew rate op amp so that the generated square waves will have vertical sides.) The circuit generates a continuous waveform by retracing the data points in one 256-byte section of the EPROM over and over.

The DAC/op amp combination produces a maximum negative voltage for a hex input of 00, a 0 V output for an input of 80, and a maximum positive voltage for an input of FF.

Table 10   EPROM Square, Triangle, and Sawtooth Waveform Data

SQUARE

Base Address	0	1	2	3	4	5	6	7	8	9	A	B	C	D	E	F
0100	FF	FF	FF	FF	FF	FF	FF	FF	FF	FF	FF	FF	FF	FF	FF	FF
0110	FF	FF	FF	FF	FF	FF	FF	FF	FF	FF	FF	FF	FF	FF	FF	FF
0120	FF	FF	FF	FF	FF	FF	FF	FF	FF	FF	FF	FF	FF	FF	FF	FF
0130	FF	FF	FF	FF	FF	FF	FF	FF	FF	FF	FF	FF	FF	FF	FF	FF
0140	FF	FF	FF	FF	FF	FF	FF	FF	FF	FF	FF	FF	FF	FF	FF	FF
0150	FF	FF	FF	FF	FF	FF	FF	FF	FF	FF	FF	FF	FF	FF	FF	FF
0160	FF	FF	FF	FF	FF	FF	FF	FF	FF	FF	FF	FF	FF	FF	FF	FF
0170	FF	FF	FF	FF	FF	FF	FF	FF	FF	FF	FF	FF	FF	FF	FF	FF
0180	00	00	00	00	00	00	00	00	00	00	00	00	00	00	00	00
0190	00	00	00	00	00	00	00	00	00	00	00	00	00	00	00	00
01A0	00	00	00	00	00	00	00	00	00	00	00	00	00	00	00	00
01B0	00	00	00	00	00	00	00	00	00	00	00	00	00	00	00	00
01C0	00	00	00	00	00	00	00	00	00	00	00	00	00	00	00	00
01D0	00	00	00	00	00	00	00	00	00	00	00	00	00	00	00	00
01E0	00	00	00	00	00	00	00	00	00	00	00	00	00	00	00	00
01F0	00	00	00	00	00	00	00	00	00	00	00	00	00	00	00	00

Table 10 EPROM Square, Triangle, and Sawtooth Waveform Data (*continued*)

TRIANGLE

Base Address	0	1	2	3	4	5	6	7	8	9	A	B	C	D	E	F
0200	80	82	84	86	88	8A	8C	8E	90	92	94	96	98	9A	9C	9E
0210	A0	A2	A4	A6	A8	AA	AC	AE	B0	B2	B4	B6	B8	BA	BC	BE
0220	C0	C2	C4	C6	C8	CA	CC	CE	D0	D2	D4	D6	D8	DA	DC	DE
0230	E0	E2	E4	E6	E8	EA	EC	EE	F0	F2	F4	F6	F8	FA	FC	FE
0240	FE	FC	FA	F8	F6	F4	F2	F0	EE	EC	EA	E8	E6	E4	E2	E0
0250	DE	DC	DA	D8	D6	D4	D2	D0	CE	CC	CA	C8	C6	C4	C2	C0
0260	BE	BC	BA	B8	B6	B4	B2	B0	AE	AC	AA	A8	A6	A4	A2	A0
0270	9E	9C	9A	98	96	94	92	90	8E	8C	8A	88	86	84	82	80
0280	7E	7C	7A	78	76	74	72	70	6E	6C	6A	68	66	64	62	60
0290	5E	5C	5A	58	56	54	52	50	4E	4C	4A	48	46	44	42	40
02A0	3E	3C	3A	38	36	34	32	30	2E	2C	2A	28	26	24	22	20
02B0	1E	1C	1A	18	16	14	12	10	0E	0C	0A	08	06	04	02	00
02C0	02	04	06	08	0A	0C	0E	10	12	14	16	18	1A	1C	1E	20
02D0	22	24	26	28	2A	2C	2E	30	32	34	36	38	3A	3C	3E	40
02E0	42	44	46	48	4A	4C	4E	50	52	54	56	58	5A	5C	5E	60
02F0	62	64	66	68	6A	6C	6E	70	72	74	76	78	7A	7C	7E	80

SAWTOOTH

Base Address	0	1	2	3	4	5	6	7	8	9	A	B	C	D	E	F
0300	00	01	02	03	04	05	06	07	08	09	0A	0B	0C	0D	0E	0F
0310	10	11	12	13	14	15	16	17	18	19	1A	1B	1C	1D	1E	1F
0320	20	21	22	23	24	25	26	27	28	29	2A	2B	2C	2D	2E	2F
0330	30	31	32	33	34	35	36	37	38	39	3A	3B	3C	3D	3E	3F
0340	40	41	42	43	44	45	46	47	48	49	4A	4B	4C	4D	4E	4F
0350	50	51	52	53	54	55	56	57	58	59	5A	5B	5C	5D	5E	5F
0360	60	61	62	63	64	65	66	67	68	69	6A	6B	6C	6D	6E	6F
0370	70	71	72	73	74	75	76	77	78	79	7A	7B	7C	7D	7E	7F
0380	80	81	82	83	84	85	86	87	88	89	8A	8B	8C	8D	8E	8F
0390	90	91	92	93	94	95	96	97	98	99	9A	9B	9C	9D	9E	9F
03A0	A0	A1	A2	A3	A4	A5	A6	A7	A8	A9	AA	AB	AC	AD	AE	AF
03B0	B0	B1	B2	B3	B4	B5	B6	B7	B8	B9	BA	BB	BC	BD	BE	BF
03C0	C0	C1	C2	C3	C4	C5	C6	C7	C8	C9	CA	CB	CC	CD	CE	CF
03D0	D0	D1	D2	D3	D4	D5	D6	D7	D8	D9	DA	DB	DC	DD	DE	DF
03E0	E0	E1	E2	E3	E4	E5	E6	E7	E8	E9	EA	EB	EC	ED	EE	EF
03F0	F0	F1	F2	F3	F4	F5	F6	F7	F8	F9	FA	FB	FC	FD	FE	FF

Switch Debouncer for the Altera UP-1 Board

The easiest way to debounce a pushbutton switch is with a NAND latch, as shown in Figure 80. The latch eliminates switch bounce by setting or resetting on the first bounce of a switch contact and ignoring further bounces. The limitation of this circuit is that the input switch must have **Form C contacts**. That is, the switch has Normally Open, Normally Closed and Common contacts. This is so that the switch resets the latch when pressed (i.e., when the Normally Open contact closes) and sets the latch when released (Normally Closed contact recloses). Each switch position activates an opposite latch function.

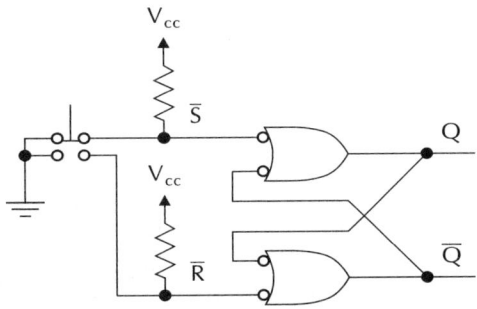

Figure 80   NAND Latch Debouncer

If the only available switch has a single set of contacts, such as the Normally Open (**Form A**) pushbuttons on the Altera UP-1 Education Board, a different debouncer circuit must be used. The circuit in Figure 81 provides a solution.

The circuit in Figure 81 is based on the same principle as a Motorola device, the MC14490 Contact Bounce Eliminator. It is adapted for use in an Altera CPLD, such as the EPM7128S or the EPF10K20 on the Altera UP-1 Education Board.

Operating Principle

The heart of the debounce circuit in Figure 81 is an Exclusive NOR gate and a 4-bit serial shift register, with active-HIGH synchronous LOAD. The XNOR gate compares the shift register serial input and output. When the input and output are *different*, the input data are serially shifted through the register. When input and output are *the same*, the binary value at the serial output is parallel-loaded back into the register.

Figure 82 shows the timing of the debouncer circuit with switch bounces on both make and break phases of the switch contact. The line labeled **4-bit delay** refers to the shift register flip-flop outputs. Pushbutton input is **PBIN**, debounced output is **PBOUT** and **CLK** is the UP-1 system clock, divided by 2^{16}. (Time values in Figure 82 are not to scale and should be disregarded.)

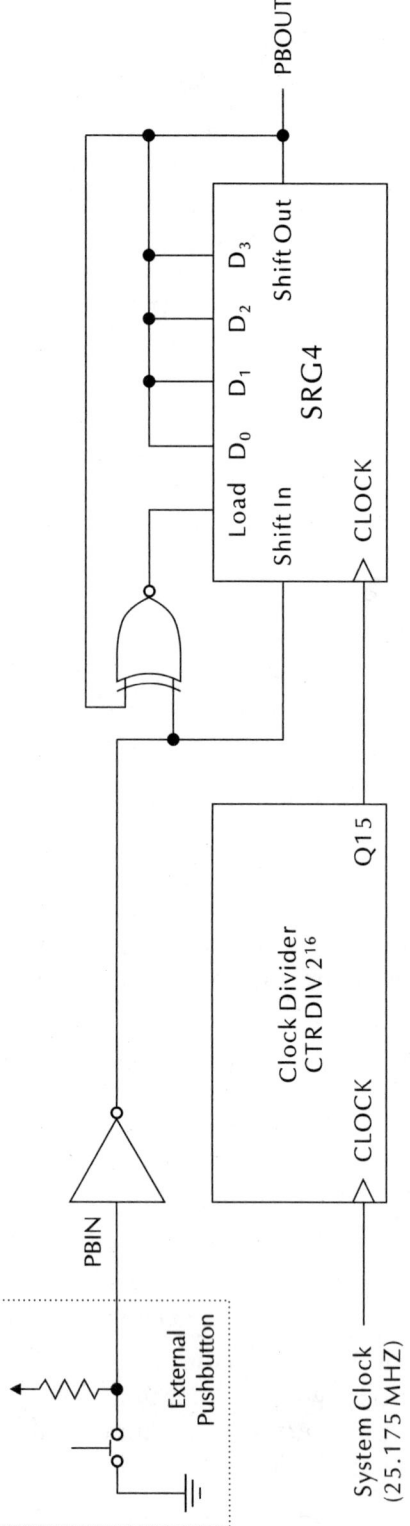

Figure 81   Switch Debouncer Based on a 4-Bit Shift Register

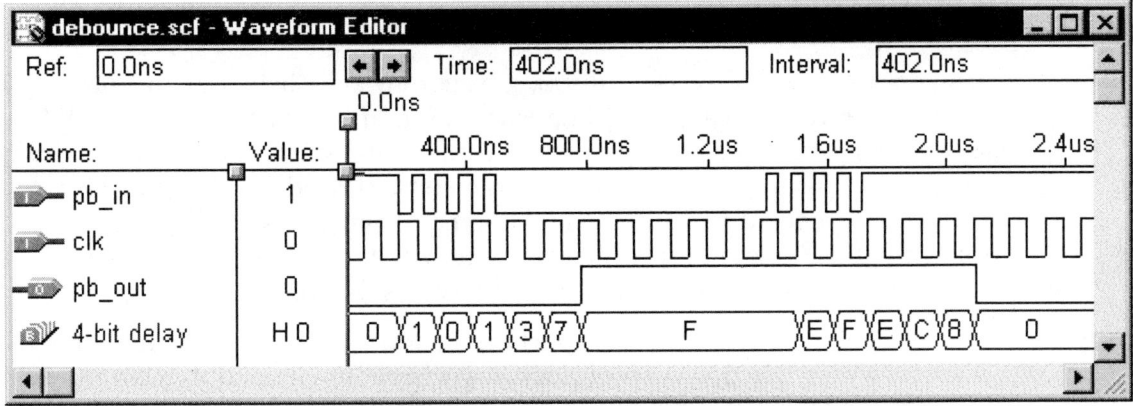

Figure 82   Simulation of a Shift Register-Based Debouncer

Assume the shift register is initially filled with 0s. The pushbutton rest state is HIGH. As shown in Figure 81, the pushbutton input value is inverted and applied to the shift register input. Before the switch is pressed, both input and output of the shift register are LOW. Since they are the same, the XNOR gate keeps the shift register in LOAD mode (LOAD input is HIGH) and the LOW output is reloaded to the register on every positive clock edge.

When the switch is pressed, it will bounce, as shown above the second, third, and fourth clock pulses on Figure 82. On the second clock pulse, PBIN is LOW. This makes the shift register input and output different, so a 1 shifted in. On the next clock pulse, PBIN has bounced HIGH again. The input and output are now the same, so the output value, 0, is loaded in parallel to all flip-flops of the shift register. On the fifth pulse, PBIN is stable at logic LOW. Since the shift register input is now HIGH and the output is LOW, the HIGH is shifted through the register. We see this by **4-bit delay** increasing in value: 0, 1, 3, 7, F, which in binary is equivalent to 0000, 0001, 0011, 0111, 1111. At this point, the input and output are now the same and the output value, 1, is parallel-loaded into the register on each clock pulse, making **PBOUT HIGH**.

A similar process occurs when the waveform goes back to the HIGH state.

To produce an output change, the shift register input and output must remain different for at least four clock pulses. This implies that the input is stable for that period of time. If the input and output are the same, this could mean one of two things. Either the input is stable and the shift register should be kept at a constant state or the input has bounced back to its previous level and the shift register should be reinitialized.

The debouncer in Figure 81 is effective for removing bounce that lasts for no more than 4 clock periods. Since switch bounce is typically about 10 ms in duration, the Altera UP-1 system clock is much too fast. The clock oscillator runs at 25.175 MHz, giving it a clock period of about 40 ns. If we divide the oscillator frequency by 65536 ($=2^{16}$) using a 16-bit counter, we obtain a clock waveform for the debouncer with a period of 2.6 ms. Four clock periods (10.2 ms) are sufficient to take care of switch bounce.

MAX+PLUS II Implementation

The text box below and on the following page shows the listing of an VHDL (VHSIC Hardware Description Language) implementation of the shift register-based debouncer. The two main components (clock-divider counter and shift register) are programmed using components from the Library of Parameterized Modules (LPM). This is a set of hardware-independent functions that allow the designer to specify a function with defined input and output ports and certain required or optional parameters. (The components are **lpm_counter** and **lpm_shiftreg**, respectively.)

An LPM module is specified by **ports** and **parameters**. A **port** is an input or output of the device, with a function such as clock, asynchronous clear, or asynchronous load. A **parameter** is a property of the block, such as **LPM_WIDTH**, which specifies how many bits its parallel input or output has. Some ports and parameters, such as **clock** and **LPM_WIDTH**, must be used in all instances of **lpm_counter**. Others, such as **aclr** and **LPM_DIRECTION**, are optional.

For example, to change the counter to any number of bits up to 255, we would simply change the parameter LPM_WIDTH to the required value. MAX+PLUS II does the work to specify the required hardware, based on the resources available in the selected CPLD. (Specifying a parameter does not necessarily guarantee that the selected module will fit into any particular device; a counter of width 255 will not fit into a device having 128 macrocells.)

To use an LPM component in a VHDL file, the ports are mapped to a set of signals in a PORT MAP and the parameters are related to the VHDL signals by a GENERIC MAP. Any interconnection between components is done with concurrent signal assignments.

For further information on LPM functions, refer to Megafunctions/LPM on the MAX+PLUS II Help menu.

```
--debounce.vhd
-- Switch Debouncer for a Form A contact, based on a 4-bit shift
-- register. Function is similar to a Motorola MC14490 Contact
-- Bounce Eliminator.
--
-- Programed by:        R. Dueck, Red River College, Winnipeg, MB

-- Use Modules from Library of Parameterized Modules (LPM):
--              LPM_SHIFTREG    (Shift Register)
--              LPM_COUNTER     (16-bit counter)

-- ieee library required to use STD_LOGIC and STD_LOGIC_VECTOR types
LIBRARY ieee;
USE ieee.std_logic_1164.ALL;
LIBRARY lpm;
USE lpm.lpm_components.ALL;

-- Define I/Os
ENTITY debounce IS
        PORT(
                clk         : IN      STD_LOGIC;
                pb_in       : IN      STD_LOGIC;
                pb_out      : OUT     STD_LOGIC);
END debounce;
```

```
ARCHITECTURE debouncer OF debounce IS
-- Internal signals required to interconnect counter and shift register
        SIGNAL srg_ser_out, srg_ser_in, srg_clk, srg_load        :STD_LOGIC;
        SIGNAL srg_data                                          :STD_LOGIC_VECTOR (3 DOWNTO 0);
        SIGNAL ctr_q                                             :STD_LOGIC_VECTOR (15 DOWNTO 0);
BEGIN
-- Instantiate 16-bit counter
        clock_divider: lpm_counter
                GENERIC MAP (LPM_WIDTH => 16)
                PORT MAP (clock    =>   clk,
                          q        =>   ctr_q(15 DOWNTO 0));

-- Instantiate 4-bit shift register
        four_bit_delay: lpm_shiftreg
                GENERIC MAP (LPM_WIDTH => 4)
                PORT MAP (shiftin   =>   srg_ser_in,
                          clock     =>   srg_clk,
                          load      =>   srg_load,
                          data      =>   srg_data(3 downto 0),
                          shiftout  =>   srg_ser_out);

-- Shift register is clocked by counter output (divides system clock by 2 ^ 16)
        srg_clk<=ctr_q(15);
-- Undebounced pushbutton input to shift register
        srg_ser_in<=not pb_in;
-- Shift register is parallel-loaded with output data if shift register
--   input and output are the same. If input and output are different,
--   data are serial-shifted.
        srg_data(3)  <=  srg_ser_out;
        srg_data(2)  <=  srg_ser_out;
        srg_data(1)  <=  srg_ser_out;
        srg_data(0)  <=  srg_ser_out;
        pb_out       <=  srg_ser_out;
        srg_load     <=  not((not pb_in) xor srg_ser_out);
END debouncer;
```

Testing the Switch Debouncer

The VHDL Design File above can be used to create a default symbol that can be used in higher-level designs, such as the two-digit hexadecimal counter shown in Figure 83.

The circuit in Figure 83 is contained in the file **2digit.gdf**. The various circuit components are included in files **sev_segv.vhd, count_8.vhd**, and **debounce.vhd**. A programming file is also included as **2digit.pof**. This file can be downloaded directly to the MAX7000S device on the Altera UP-1 board.

You can create your own programming or configuration file by compiling the design in Figure 83, using the supplied VHDL files. To use these files, copy them to a new subfolder in your MAX+PLUS II working folder. Open MAX+PLUS II and create a default symbol for each VHDL file. Add your new folder to the list of user libraries.

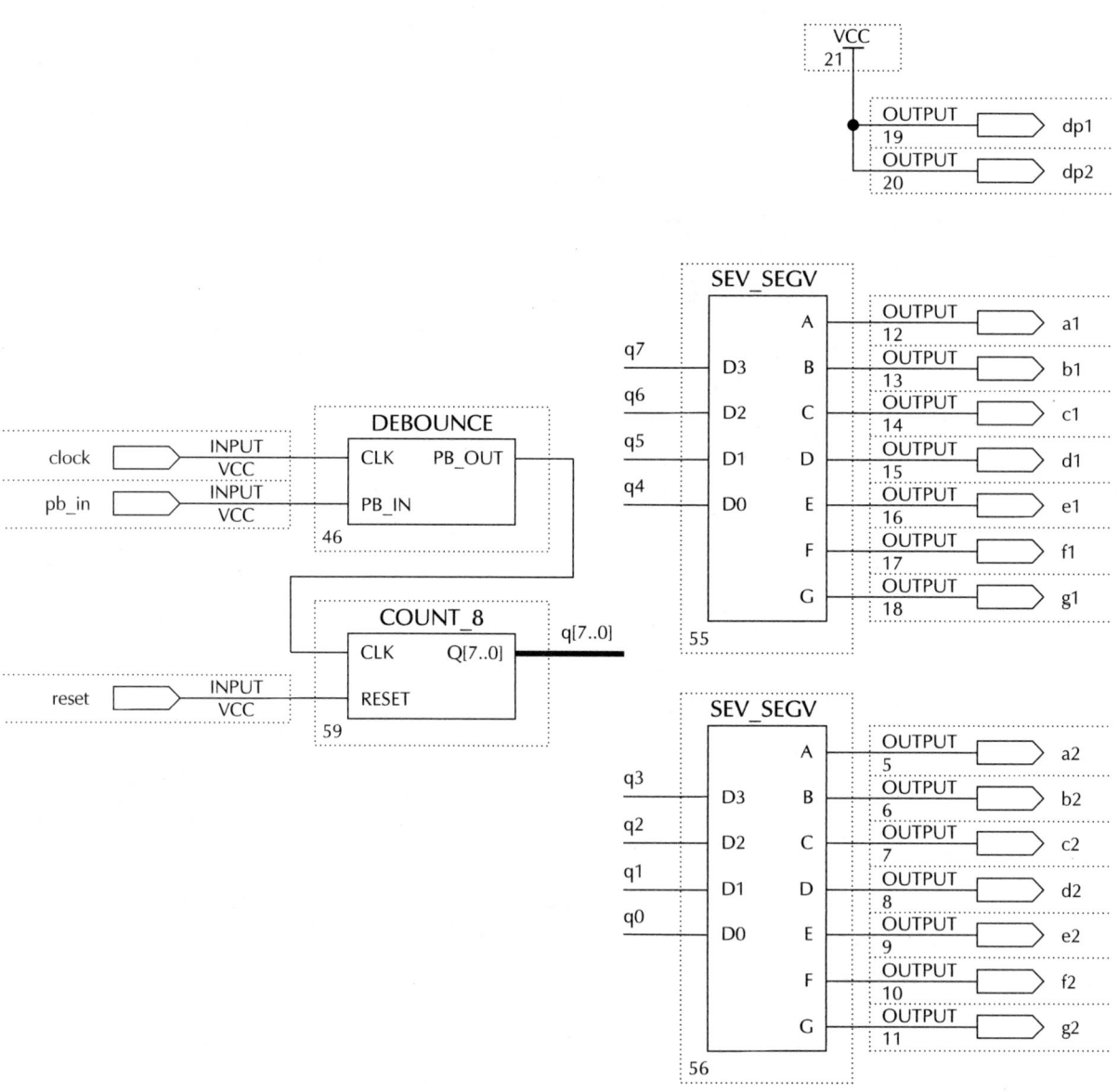

Figure 83 Two-Digit Hexadecimal Counter Showing Use of Switch Debouncer

Open the highest-level design (**2digit.gdf**) and add pin numbers if necessary. Pin assignments are shown in Table 12 for the EPM7128S (MAX7000S) device and in Table 13 for the EPF10K20 (FLEX10K) device.

Table 12
EPM7128S Pin Assignments for Figure 5

Function	Pin	Function	Pin
CLK	83	PBIN	51
RESET	52		
a1	58	a2	69
b1	60	b2	70
c1	61	c2	73
d1	63	d2	74
e1	64	e2	76
f1	65	f2	75
g1	67	g2	77
dp1	68	dp2	79

Table 13
EPF10K20 Pin Assignments for Figure 5

Function	Pin	Function	Pin
CLK	91	PBIN	28
RESET	29		
a1	6	a2	17
b1	7	b2	18
c1	8	c2	19
d1	9	d2	20
e1	11	e2	21
f1	12	f2	23
g1	13	g2	24
dp1	14	dp2	25

Compile the file **2digit.gdf** and download it to the MAX7000S device (EPM7128SLC84-7) on the Altera UP-1 board. Connect a wire from one of the pushbuttons (CLK) to Pin 51 on the MAX prototyping header and the other pushbutton (RESET) to Pin 52. Alternatively, compile the file **2digflex.gdf** and download it to the

FLEX10K20 device. Pushbuttons and seven-segment outputs are hardwired to the FLEX device.

In each case, the seven-segment outputs on the counter should advance by one, counting in hexadecimal, each time the CLK switch is pressed. The corresponding counter should reset to 00 when a RESET switch is pushed.